7·26·78

CLIPPER 806

The Anatomy of an Air Disaster

John Godson

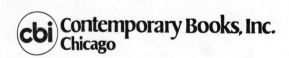
Contemporary Books, Inc.
Chicago

Library of Congress Cataloging in Publication Data

Godson, John.
　Clipper 806.

　　1. Aeronautics—Accidents—1974.　2. Boeing 707
(Transport planes)　I. Title.
TL553.5.G6　　1978　　　996'.13　　　77-23693
ISBN 0-8092-8100-7

Copyright © 1978 by John Godson
All rights reserved.
Published by Contemporary Books, Inc.
180 North Michigan Avenue, Chicago, Illinois 60601
Manufactured in the United States of America
Library of Congress Catalog Card Number: 77-23693
International Standard Book Number: 0-8092-8100-7

Published simultaneously in Canada by
Beaverbooks
953 Dillingham Road
Pickering, Ontario L1W 1Z7
Canada

Other books by the author

Unsafe at Any Height
Runway
Papa India—The Trident Tragedy
Design by Accident
The Airline Pilot
An Eye for an Eye
The Rise and Fall of the DC-10

Foreword

This story of a group of travelers needs only, by way of introduction, the names of those taking part. Following, then, is the passenger manifest for Pan American World Airways' Flight 806—scheduled from Auckland, New Zealand, to Pago Pago, American Samoa—that departed on the evening of Wednesday, January 30, 1974.

I. Ahkam
A. Amerperosa
B. Atkinson
E. Bennett
F. Bennett
H. J. Blaesi
J. G. Blaesi
J. C. Bonham
I. Brewster
N. M. Brice

H. M. Cann
R. D. Cann
L. Chernoff
S. Chernoff
A. Chongwong
A. Coldicutt
A. L. Coldicutt
J. Coldicutt
C. Coman
C. Cosby

G. H. Cox
S. D. Chrichton
C. Culbertson
W. Dearmond
G. Debroff
R. Debroff
A. J. Dilby
B. Dunsby
C. Dunsby
G. R. Dunsby
N. Dunsby
W. B. Evans
K. Fidow
J. W. Folder
A. Forbes
A. P. Gates
M. Gates
B. Giles
R. T. Giles
V. Green
T. Helberg
M. Henwood
A. Hill
R. F. Hill
T. Iafeta
T. Isale
D. Killips
L. Kinderman
R. Kinderman
S. Leleva
E. Lewis
J. Lewis
M. J. Lewis
P. J. Lewis
W. Lewis
M. M. Luikui

A. Masiu
A. Matuatia
S. Matuatia
D. McKay
J. Merrill
M. P. Merrill
E. Muenster
S. Munro
G. Orton
J. Orton
N. Orton
S. Orton
V. Orton
D. L. Phillips
L. Phillips
J. Price
M. Quinn
B. Redford
J. Redford
A. Rode
S. Sauao
R. A. Smith
T. Smith
C. Sweeten
R. Sweeten
P. Tauiela
M. Taylor
T. Toilolo
S. Tsai
B. Unise
G. van Boxtel
D. Weyland
E. Weyland
B. Willetts
G. Winteringer

The ninety-one passengers included forty-one male adults, thirty-eight female adults and twelve children. There were fifteen families.

The rostered crew for Flight 806 consisted of

    Captain Leroy A. Petersen
    First Officer Richard V. Gaines
    Third Officer James S. Phillips
    Flight Engineer Gerry W. Green
    Senior Purser Gorda Rupp
    Junior Purser Elizabeth Givens
    Stewardess Patricia Reilly
    Stewardess Gloria Olsen
    Stewardess Kinuko Seki
    Stewardess Yvonne Cotte

AUTHOR'S NOTES

On Pronunciation: Two tricky place names that appear regularly in this book are not pronounced as they are spelled. For those wishing to be ultra-correct the names are Nadi (*Nan-dee*) and, of course, Pago Pago (*Pongo Pongo*).

On times: All times are local, based on the twenty-four-hour clock. Hours and minutes are indicated by the letter *h* and, where necessary, this letter is followed by the seconds.

On timings: The events in the escape sequence are related in chronological order. The times given, however, are approximations, based on the accounts of those taking part, on eyewitnesses, as well as on "dead timings" from the aircraft's flight recorders and the airport's central recording apparatus.

Prelude

Civil air accidents actually are rare. Ninety-nine percent of all flights end without incident. It is the remaining one percent, however, that is cause for concern—not because accidents happen, but because of the reasons they are allowed to take place.

Are these accidents examples of history repeating itself—something we have been told does not occur?

Above all, are the disasters themselves—together with their associated pain and suffering, destruction and death—avoidable?

This is an account of one such accident, a disaster in which some world-renowned names in aviation, a typical group of international passengers, and a tiny, tropical, mid-Pacific island, came tragically together.

1

When Clipper 806 arrived in Auckland, New Zealand, Captain Leroy A. "Pete" Petersen was relaxing over a cup of coffee in the Pan Am offices at the airport with the other members of his crew: First Officer Richard V. Gaines, Third Officer James S. Phillips and Flight Engineer Gerry W. Green. There was an unusual amount of laughter coming from the group because Dick Gaines, who was suffering from laryngitis, had lost his voice.

Finally, after *all* the reasons for the first officer's untimely silence had been discussed, Captain Petersen suggested that, under the circumstances, it might be better for Gaines to trade jobs with Phillips for the coming journey.

Elsewhere in the terminal building the members of the cabin staff were conducting a preparatory briefing. Their chief purser, Gorda Rupp, was, at 34, the eldest. Her assistants, in order of seniority, were Elizabeth Givens, Gloria Olsen, Patricia Reilly, Kinuko Seki and Yvonne Cotte who, at 23, was the youngest.

The Boeing 707-321B was being operated as Pan Am's normal Wednesday evening flight from Sydney, Australia, to Auckland, and then on to Pago Pago, American Samoa; Honolulu, Hawaii; and finally San Francisco, where it terminated. The Clipper had arrived several minutes early in Auckland, and the crew that had brought her in signed off, handing responsibility over to the Petersen team, who were to fly her as far as Honolulu.

After taking their places in the cockpit, Petersen and his staff inspected the log. It indicated that there had been no major difficulties with the ship on its last few journeys, other than two "bubbles" in the captain's side window—one in the upper-left-hand and the other in the lower-right-hand corners. These were noted as being "within the limits" allowed for fenestration manufacture defects. There were five other minor problems which, during the stopover, were being remedied.

Petersen and his team then conferred with the local Pan American "Red Cap"—the director of loading operations—who told them they should be taking off on time, at 2000h, with ninety-one passengers—ten in the first-class section and eighty-one in economy—and that there would be a large amount of cargo. The Red Cap showed Petersen the planned loading manifest:

1 box—aircraft parts—weight 18 lbs.
1 box—camera parts—weight 2 lbs.
1 box—personal effects—weight 13 lbs.
1 box—aircraft parts—weight 7 lbs.
1 box—automotive parts—weight 22 lbs.
1 box—printed matter—weight 15 lbs.
1 cot and mattress—weight 53 lbs.
1 manitar earth unit—weight 13 lbs.
1 box—cine film—weight 18 lbs.
4 boxes—tubes, pulleys, and sheaves—weight 214 lbs.
3 boxes (of a consignment of 78)—fruit and vegs.—weight 33 lbs.
33 boxes—vegetables—weight 723 lbs.

1 packet—stainless trays—weight 7 lbs.
1 box—filler catalyst—weight 20 lbs.
29 packets—foodstuffs—weight 1,078 lbs.
94 packets—frozen cakes—weight 2,028 lbs.
11 packets—foodstuffs (seafood)—weight 304 lbs.
1 packet—marine components—weight 46 lbs.

Captain Petersen had more than just a passing interest in the weight of the materials his plane would have to carry. January, 1974, was the middle of the misnamed "fuel crisis," and Pan Am pilots had been instructed to load on board as much kerosene as they possibly could take, whenever they could get it, in case they should encounter a shortage at the next stopover point.

Petersen decided to carry 14,500 gallons of fuel in the tanks, bringing the overall heaviness of the 707 to just 600 pounds under its maximum permissible take-off limit of 295,500 pounds. The flight to Pago Pago should use about 6,075 gallons, he calculated. If he were unable to land in Samoa, he would need an extra 3,675 gallons to get back to Nadi on Fiji. But if there were no difficulties, the 707 would arrive at Pago Pago with 8,425 gallons on board, just about enough to push it to its next stop, Honolulu. Whatever happened, the plane was heavily laden with fuel.

While the kerosene and freight loading continued, the first passenger call went out over the speakers in the Auckland International Airport terminal. Those with tickets for Clipper 806 were hustled through immigration and customs, then had to wait in the transit lounge for a few minutes while the cabin staff checked the food and refreshments on board and did a bit of tidying up.

Finally the boarding announcement was made. The ninety-one travelers emplaned.

Among the last to enter the cabin were Mr. and Mrs. Roger Cann. It was their first flight, and they didn't quite know what to expect. In preparation for the event, Cann had read several military books on aviation.

Roger D. Cann was twenty-three years old. As a boy he,

an older brother, and a younger sister had lived with their family in Hamilton, New Zealand, where their father was a company director. In 1969, Roger had left to attend the Auckland Law School.

Heather M. Cann was just six days short of turning twenty-two. The youngest of three children of an Auckland builder, she taught dancing in a local studio.

In early 1973, Roger, fresh from completing his bachelor of law degree, had started work with a firm of Auckland attorneys while studying for his Bar examinations. He had met Heather three years previously, but it was not until July, 1973, that they were engaged. They were married on September 15 of that year, and moved into a small flat in Auckland. It was a difficult period for them. Heather was involved in her classes most nights; Roger worked long hours for only about $U.S.45 a week.

By November they were both fed up. They decided to travel and see some of the world. Roger's brother and his wife were in San Francisco; that would be an ideal way to start. In exchanges of letters they formulated a plan. Roger and Heather would fly to the United States and then, using the brother's automobile and camping equipment, the four of them would tour Mexico and the American West before Roger and Heather left for Europe.

With the organizing part roughly settled, they booked a tour of Scandinavia and Russia which would begin the first week in April. After that, they decided, they would spend the summer in hot, sunny southern Europe.

In December, Roger resigned from the law firm. He immediately took another job as a haymaker working seven days a week from daylight until dark, trying desperately to earn extra money for the trip.

At the end of January they returned to Auckland. That Wednesday, January 30, they arrived at the airport and waited impatiently with family and friends for their flight to be called. When they boarded the airliner they looked for seats 16-A and 16-B—the ones they had been assigned—on the left-hand side just behind the wing. They took their places and, following the lighted overhead instructors, fastened their seat belts.

A short while later the engines started. As the plane taxied away from the loading gate, the music on the loudspeakers was suddenly cut and a semi-audible voice announced that they would be shown the normal emergency procedures for all long-distance overwater flights.

Heather watched intently, trying to concentrate on the stewardess conducting the demonstration, who was standing in the passageway between the first-class and tourist-class sections. Heather had some difficulty seeing above the heads of the other passengers and the seat backs. The only information she managed to absorb was that the life jacket had to be put on over her head. She checked to make sure that the vest was under her seat.

Roger also paid close attention to the emergency briefing. He was very safety conscious and, perhaps because this was his introduction to the world of flying, he later remembered details which he might otherwise have forgotten. It occurred to him that the demonstration seemed very casual. He would have preferred something a little more dramatic. In any case, he felt he had learned nothing that he had not known before.

As the aircraft taxied out toward the runway, the briefing concluded with the request that passengers take a few minutes to read the *Emergency Instructions* folders in the seat pockets in front of them.

A few rows ahead of the Canns on the opposite side of the cabin was Richard A. Smith, a short, stocky, athletic man who was sports director of Williams Air Force Base and owner of the Dick Smith Swim Gym in Phoenix, Arizona. Smith had been in the United States Air Force and a colonel in the Air Force Reserve for some thirty years but had no piloting experience.

Dick Smith was then fifty-seven years old, and his whole life had been devoted to sports. He had graduated from the

University of Southern California in 1940 and since then had built an international reputation as a diving coach. He had been in charge of the American diving teams in both the 1964 Tokyo and 1968 Mexico City Olympic Games, and was a member of the U.S. Olympics Committee as well as president of the World Diving Coaches' Association.

"Discipline and dedication," he insists, ". . . are absolute necessities for record-beating sportsmen. My job is to teach divers how to overcome fear in panic situations." Little did he know that he was soon to face such a test himself.

In 1973, Smith had been invited to Australia to coach some local divers in preparation for the national championships being held in Adelaide. He was there for about a month. He was then persuaded to stop off in New Zealand on the way home and have a look at the standards of their diving team. The New Zealanders were so impressed by him that they asked whether he would help prepare their divers for the British Commonwealth Games which were to be held in Christchurch, New Zealand, in January 1974. Smith agreed.

He flew back to Auckland just after Christmas 1973. That gave him three weeks of solid training. While the games were in progress, he learned that he was to be presented with the Fred Katey Award—the most coveted swimming trophy in the U.S.—and that the ceremony would take place in Los Angeles on February 2. It was indeed a great honor, but the timing complicated things. He had planned to return to America via Jakarta, Singapore, Bangkok, Manila, and Tokyo, calling on some of the local divers along the way. The award, however, was more important. He would have to change his plans.

He contacted a Christchurch travel agent and was told that if he left two days before the end of the games, he could make it back to Los Angeles in time. Because of the length of the fourteen-hour flight, he wanted to stop off somewhere en route to stretch and rest; and this meant either Pago Pago or Honolulu. He decided on the former, primarily because it would give him a chance to visit friends—the governor of Western Samoa and a family to whom he had once been introduced.

Smith flew to Auckland on the afternoon of Wednesday, January 30, and presented himself at the Pan Am check-in area, where he was to collect his new ticket. He then walked over to a bank to change some money. As he was strolling back toward the check-in desk he was called by the clerk. Another man, surrounded by parcels and hand luggage, was standing there in a state of confusion. The receptionist asked him if he was Mr. Richard A. Smith. He replied that he was. Thank goodness, said the clerk. The other man was also a Mr. Smith, Thomas Smith from Sydney. They had been given each other's tickets. A quick exchange solved that problem.

While the two Smiths were chatting, a couple of young Americans carrying backpacks and other camping paraphernalia arrived at the counter and asked the Pan Am employee if they could be given seats away from one another. It seemed an odd request; after they left, the two Smiths and the clerk speculated about the reason for it. They finally concluded it must have been because of the amount of material each was carrying. If they were apart, they would be more likely to have room to stretch out and sleep during the flight.

When Smith boarded the plane, he noticed that one of the backpackers was in seat 21-F, directly in front of him, while the other was in the next seat forward, 22-F.

As the plane taxied away from the boarding gate, the stewardess's briefing began. Smith had heard these things so many times before that he preferred watching the activities outside on the apron. But when he heard the stewardess mention the emergency folder, he looked for it among the assortment of material in the seat pocket in front of him. He usually read such folders to check the positions of the exits in his vicinity and to discover how to open them. He accidentally pulled out one of the Pan Am magazines and, amidst all the bright advertisements and route maps, found what he thought he was searching for.

The booklet he had chosen, even though a section of it dealt with what loosely were termed "Emergency Instructions," was not the one the stewardess had referred to. Instead, it was a Pan Am brochure designed to promote other trips on the airline's services. The evacuation details it contained, therefore, were *very* limited.

The fact that he had picked up the wrong document did not even occur to him. After looking at hundreds of different cards over the years, he was of the opinion that none of them was very clear nor very complete.

Roger Cann had found two booklets in the pocket in front of him. One was evidently a publicity thing. The second was a narrow, long, blue-shaded folder. He withdrew this and discovered that it was written in many different languages—eight, to be exact. His first impression was that he would get little information from it. He put it to one side and pulled out the same Pan Am advertising pamphlet Smith had discovered. Somewhere in the middle he, too, found the escape instructions, which he examined in detail. There was a diagram showing the position of the over-wing emergency exit windows. He recalled that during the stewardesses' demonstration one had gestured in their general direction, without pointing directly at them. Now that he had the plan, Cann made it his business to learn exactly where they were.

Heather Cann also extracted the blue folder from the seat pocket. It seemed to have explicit information on the life jackets, but that was the only thing in it that struck her as being clear. She also noted the page dealing with the exits, and tried to understand how to operate them from reading the details. She was of the opinion that if the handle was turned, the doors would magically open. She had no idea which way they would swing. She presumed outwards.

She returned the folder to its former place. Looking

around, she saw that no one in their immediate vicinity other than her husband seemed to be interested in it at all. The whole briefing had been quite casual, she later said, as if nobody would remember it anyway.

Now, as Captain Petersen steered Clipper 806 toward the take-off runway, Heather Cann noticed that there seemed to be an enormous amount of luggage spread around the passenger cabin. She recalled that when they had boarded the plane, people brought a lot of things with them. She could see a number of kit bags on the overhead hat racks. She did not know it was against regulations to place them there. She did realize that, because of their bulk, it would have been impossible to stow them beneath the seats.

The Boeing 707 suddenly came to a halt. It was too dark outside for the Canns to see what the problem was. Actually, nothing was wrong. They had simply arrived at the holding area just off the end of the strip, and air traffic control had instructed the pilots to wait a few minutes for runway clearance.

Captain Petersen sat tensely in his seat. He still became nervous at the start of a journey, even after the 17,414 hours of flying he had clocked during his career.

"Pete" Petersen, like the others in his crew, was based at Pan Am's Los Angeles headquarters. He had been with the company since March 1951, had worked his way up through the subordinate ranks, and in November 1960, passed his Boeing 707 training as a reserve co-pilot/navigator. He was upgraded to master co-pilot for the 707 fleet in July 1965. His next step on the ladder, to captain, was made in November 1967.

Leroy Petersen had had his share of medical problems during the years. Pan Am's records show that he had an hemorrhoidectomy on February 26, 1968; a vagatomy and pyloroplasty on October 9, 1968; a hernia operation in July 1970; a kidney stone removed on October 1, 1970; and had recently been sick—from September 5, 1973, to January 15, 1974—with a gastrointestinal ailment. Because of this last illness, he had had to submit to a series of tests showing that during the time off he had not forgotten what flying was all about. "By the end of the period," his examiner's report states, "Captain Petersen was doing very good work, including three engined flight-director instrument landing system approaches to Category II minima."

His normal A-phase aviation check was completed on January 18, 1974, less than two weeks prior to the present take-off. The record contains the notation that he exhibited a good knowledge of systems and procedures and that his "simulator" work was "very well done throughout." In order to requalify in the aircraft following his long layoff, he then made three complete flight sequences on January 19, with the company's inquisitors sitting behind him.

Since that time, he had flown for just thirty-five hours, and over the last eight days he and the others in his crew had worked a unified roster pattern—one that was scheduled to keep them, as a team, away from their homes in California until early in February.

Petersen was fifty-two. In 1941 he had enlisted in the U.S. Navy's flying school and on qualifying had become a test pilot. At war's end, after being discharged, he had attended the University of Utah, graduating in 1948 with a Bachelor of Science degree in business administration and marketing. In March 1947, he married Betty Joyce, two years younger than he. They had two daughters, Patricia Anne, born in April, 1950, and Mary Anne, born nearly ten years later. Pat was currently studying for a B.A. at Westminster College, and Mary was attending high school in Salt Lake City.

In between his flying duties with Pan Am, Petersen dabbled in property management and investment. He was also a part-time insurance salesman.

Now, as they awaited the controller's clearance to move onto the Auckland runway, Petersen adjusted his glasses and, turning to Jim Phillips, his acting co-pilot, suggested that he might like to effect the take-off.

James Sheridan Phillips was eight years younger than Petersen. After attending the College of Marin he had enrolled, in 1950, in the University of California at Berkeley, intending to become a veterinarian. But, in 1953, he apparently changed his mind and enlisted in the military, eventually becoming a test pilot in the Marine Corps. Soon after he was released in 1958, he married. The union did not last, however; he was divorced six months later.

In 1962 he married Shelbyjean. Their only child, Dean, was born in July 1964. The father and son became very close, sharing interests in boating and waterskiing. Young Dean spent many happy hours with his dad out on the coastal waters of the Pacific.

Jim Phillips started working for Pan Am in April 1966, and after less than ten months had completed his initial Boeing training, qualifying him as a reserve copilot/navigator. Up to now, about ninety percent of the 5,208 hours he had flown had been in 707s.

Phillips had passed his last A-phase flight checks on November 14, 1973. His examiner stated in his report, "Good work. Should rate in six hours." His most recent simulator practice had been conducted on May 7, 1973, on which occasion he had received the following comments: "All areas at good level of proficiency. OK for line landings," meaning that Pan Am was quite happy to have him actually fly their aircraft.

Unlike his skipper, Phillips had absolutely no medical record at all.

The acting co-pilot nodded assent to Petersen—certainly he would effect the lift-off. This may have come as a disappointment to the person who had originally been rostered as co-pilot but who had lost his voice a few hours before. The silent one had been relegated to a position just behind the captain, where he was to keep a watching brief on the actions of the two operating flyers.

Richard Victor Gaines was thirty-seven years old. He had attended West Point for one year, then transferred to three other colleges in succession: the University of Colorado, East Carolina College, and the University of Maryland. Yet he never obtained a degree. Instead he ended up as a pilot in the U.S. Marines, joining Pan Am in August 1964, after his discharge. Three months later he had completed his initial Boeing 707 reserve co-pilot/navigator tests, and in June 1967, he was upgraded to a master co-pilot's position.

On January 18, 1974, the same day Captain Petersen was undergoing his checks, Gaines passed his A-phase flight exams. Over the preceding twelve months he had flown into Pago Pago airport a dozen times.

Eight months before he started with Pan American World Airways, Dick Gaines had married. Both he and his wife, Myrlyn, had wide-ranging interests, including numismatics and collecting military memorabilia. Dick's hobbies included learning to play, studying the history of, and composing music for the Flamenco guitar. On top of all this, he dabbled in the commodities market and was quite successful until the bottom began to drop out of it in the early 1970s. He was also involved, to a lesser degree, in real estate construction. Myrlyn, in fact, was supervising the building of a four-bedroom house they hoped to rent out for investment purposes. They had, by then, a three-year-old son, Richard V. Gaines II.

The other crew member, Flight Engineer Gerry Ward Green, was also thirty-seven. He had been with Pan Am since April 1967, but had not obtained his engineer's qualifications until July 1973.

Green had attended Washington State University until 1954, then transferred to the University of Oregon where, in 1961, he had received a Bachelor of Science degree in mathematics. The following year he had become a pilot in the U.S. Air Force.

In 1963 he had married, and in January 1965, his son Steven had been born. But tragedy struck. In January 1973, his wife suddenly died. He was left with a large house, a job that continually took him long distances from home, and an eight-year-old to look after. Though nobody understood why, Green had made arrangements to sell his home and had signed the contract just four days previously, on January 26, 1974.

Like the other three in Clipper 806's cockpit, Green had been working the same unified shift pattern. In the twenty-four hours prior to the Auckland departure, they had been off duty for nineteen and a quarter of them. Now they sat and waited for permission from the control tower to start their next journey. Soon it would be brakes-off time, and away.

2

In the early afternoon of January 30, 1974, employees of Consolidated Chemicals Ltd. in Auckland were rushing to finish an order. It had been placed by the Government of American Samoa and was for 200 twenty-cc. plastic tubes containing what the manufacturer described as a "filler catalyst." Though it is not known why it was wanted, the so-called filler catalyst was the highly explosive methyl ethyl ketone peroxide, more commonly tagged as MEK peroxide.

Methyl ethyl ketone, an alcohol, is exceptionally inflammable. It smells something like acetone, and has a "flash-point" (the temperature above which it gives off ignitable fumes and can explode) of thirty-three degrees Fahrenheit—just above freezing. It is most often used as a solvent. It is also employed in making surface coatings for buildings and in the fabrication of clear synthetic resins—popularly, polyvinyl chloride. With the addition of peroxide, the "flash-point" temperature is increased to 124 degrees (still very combustible), and the resulting mixture is used mainly in the textile industry, as a bleach. MEK peroxide is also highly narcotic if the fumes are inhaled and causes extreme irritation to the eyes and mucous membranes. It is, therefore, on the airlines' *Restricted Articles* list because of its combustibility and its effects on humans should its containers break; above all, it must be handled very carefully.

The guidelines for the conveyance of such substances between countries has been laid down by the International Air Transport Association (I.A.T.A.), the carriers' price-and-conditions fixing cartel, which manages to regulate everywhere despite its being an illegal organization in a number of the world's territories. These precepts state that the most that may be packed into any one container is one pound, or one pint, or just twenty-eight of the twenty-cc. tubes. Compatible plastic ampules of not over five cc. must be used (Consolidated was packing them in capsules four times as big), and they must be cushioned with noncombustible, absorbent material. These, in turn, are to be loaded, according to the rules, into fiberboard containers, and no more than twenty-four of the receptacles are to be placed into any one crate. The maximum quantity in an individual crate must not exceed two pounds, or two pints.

The 200 small plastic vials (totaling twenty-two pints) had been assembled. Consolidated Chemicals Ltd. was fully aware that the contents were listed in the cartel's *Restricted Articles Manual,* but because the MEK peroxide had been diluted with hydroquinone, increasing the "flash-point" to 180 degrees, it was considered "safe enough." Consolidated believed that the criterion for hazardous materials was solely their explodability.

The small tubes were packed into four tins which were then filled with perlite (a noncombustible, absorbent substance). The cans were sealed and placed into a single fiberboard carton marked "Consignment to—The Government of American Samoa, Pago Pago Airport—Filler Catalyst," and the crate was handed over to the local I.A.T.A. certified forwarding agent, Meadows Air Freight Ltd. A Meadows clerk wrote out airwaybill number 026/3397 9525, describing the contents of the consignment as "Filler Catalyst—(non hazardas [sic])," and had it delivered to the P.A.A. freight office at Auckland International Airport.

It was only hours later that employees of Pan Am loaded the restricted material on board Clipper 806, soon bound for Pago Pago. MEK peroxide is *not* a "filler catalyst," whatever that is supposed to mean.

Pan American World Airways had been caught in a violation of I.A.T.A. rules regarding chemical shipments before, just eighty-eight days previously, in fact.

On that occasion the plane—a freight-only model of the Boeing 707—had been standing at its Kennedy Airport cargo terminal on the morning of November 3, 1973, while the Red Cap (the director of loading operations) was in the cockpit chatting with the crew—Captain John J. Zammett, Co-pilot Gene W. Ritter, and Engineer David X. Melvin. They were preparing for Flight 160, from New York to Frankfurt, Germany, with a stopover at Prestwick, Scotland.

The Red Cap knew them well. Soon after he entered the airplane and recognized them, he placed the loading documents, which he had brought with him for Captain Zammett to sign, on top of a dispatch box just outside the cockpit door. The conversation became so animated that when it was time for take-off there was a sudden scramble to close and lock the forward exit, start the engines, and get under way.

The forgotten cargo manifests remained on the dispatch box, unnoticed.

The first part of the journey went well. They took to the air at 0824h, nineteen minutes late. The pilot was climbing up to his planned cruising height of 33,000 feet, when the New York air traffic controller radioed him requesting that he stay at 31,000 feet because of other aircraft. The crew reported back at 0850h, saying that they were at that flight level.

At 0904h22, the special Pan American Operations frequency came to life (radio conversations are designated by asterisks):

          OPERATIONS: *One Six Zero. Pan Ops. Go ahead.
0904h24 PILOT: *Ah, yes sir. We have . . . ah . . . an accumulation of smoke in the lower forty-one [the equipment bay area under the cockpit] and we're gonna go back to Boston. Do you want us back at Boston or New York?
0904h34 OPERATIONS: *Ah . . . stand by One Sixty. We'll find out.

The crew members' conversations continued as they watched white smoke puffing under the door that separated the flight deck from the freight-filled cabin:

0904h43 PILOT: New York is not that much further on, so we can just go ahead back.
          CO-PILOT: Do you think . . . Do you wanna go to New York?
          PILOT: I asked him, where did he want us now?
          PILOT: Put New York [frequency for the distance measuring equipment] on your[s] [radio], and see how far we are from it.
          CO-PILOT: It won't show.
0905h07 PILOT: Dave!
          ENGINEER: Yeah?
          PILOT: You don't think you should get down there [into the lower equipment bay] and spot that, huh?
          ENGINEER: I can't get around down there at all.
0905h15 ENGINEER: I don't see any reason up here why that son-of-a-bitch . . . It shoulda popped a

              [circuit] breaker by now. It oughta short out somewhere.
        CO-PILOT: Can we increase the airflow [in the cockpit] so we can get rid of some of the smoke through the outflow valves?
        PILOT: Yeah.
0905h38 PILOT: Just stick your head down there and see if it's still coming up.

    The crew seemed to think that the smoke was originating from an electrical fire, possibly in the electronics bay beneath the rear of the cockpit where the engineer was seated. With the aircraft on automatic, the three of them were fascinated by the white cloud seeping under the door, even though it was, at this time, becoming thicker.
    They suddenly realized that their Operations Room back at Kennedy Airport was very slow:

0905h49 PILOT: *Pan Ops from Clipper One Sixty.
        OPERATIONS: *Sixty. Go ahead.
        PILOT: *Ah . . . Did you get that message? Do you want us to come back to New York or go to Boston?
0905h59 OPERATIONS: *One Sixty. They're checking on that right now. Copied you've got an accumulation of smoke in your lower forty-one. They're . . . ah . . . finding out where they would like you.

    The problem was that if the fire was a technical one as they thought, Operations would either need a full engineering crew to remedy the cause, or they would need another aircraft to which everything on board would have to be transferred. Meanwhile, from the talk on the other radio receiver tuned in to air traffic control, things carried on relatively normally:

0906h03 BOSTON: *Clipper One Sixty. Contact Montreal Center [on radio frequency] one three two point . . . ah . . . three five, and make your request to them.
PILOT: *OK. We'll stand by.
OPERATIONS: *Ah . . . One Sixty. They say come back to New York, and . . . ah . . . when you get a moment, you can give us a good ETA [Estimated Time of Arrival] for New York.
0906h17 PILOT: *Stand by. We'll just get our . . . ah . . . [air traffic control] routing back to New York firstly.

The pilot then instructed the co-pilot to call the Boston controllers, telling them that they wanted to return to New York without bothering with Montreal:

0906h23 CO-PILOT: *Ah . . . Clipper One Sixty. Requesting present position direct to New York. *Direct New York* at this time.
0906h32 BOSTON: *Clipper One Sixty. Contact Montreal Center, one three two point three five, and . . . Montreal Ce . . . ah . . . Montreal Center.
0906h41 CO-PILOT: *Roger. Roger.
PILOT: Bastards.
0906h53 CO-PILOT: *Montreal Center good afternoon, it's Clipper One Sixty.
0907h07 CO-PILOT: *Montreal Center—Clipper One Six Zero.

While the co-pilot was attempting to contact Montreal for a clearance to turn about, the other two crew members were occupied with the smoke:

0907h15 ENGINEER: This fucker sure is comin', John.

0907h20	ENGINEER:	Oh, Christ. Lemme see if I can shut this ... ah ... blower off.
0907h30	ENGINEER:	I'm gonna raise the cabin [pressurization] up.
	CO-PILOT:	Did you get the Montreal [radio] frequency? It's the only frequency I didn't write down.
	ENGINEER:	Up to twelve? [Requesting whether he should lower the cabin pressure to the same as it is outside at 12,000 feet altitude.]
	PILOT:	Let's see.
0908h24	CO-PILOT:	*Montreal Center. Clipper One Sixty.
0908h31	ENGINEER:	Could be a bleed....
	PILOT:	All right.
	ENGINEER:	... and try and get some air into this fucking place.
	PILOT:	Oh, shit.
	CO-PILOT:	*Montreal Center. Clipper One Sixty.
0908h45	MONTREAL:	*Clipper One Six Zero. Montreal. Squawk ... ident, and say your altitude.
	PILOT:	Tell 'em we wanna go to fucking New York ... DIRECT.
0908h48	CO-PILOT:	*Clipper One Sixty. Level at three one zero [31,000 feet altitude]. And we wanna go right back to Kennedy at this time.
	MONTREAL:	*Clipper One Six Zero. Roger. Turn right, heading one eight zero [degrees].
	CO-PILOT:	*Right turn to one eight zero. Thank you.
	MONTREAL:	*And Clipper One Six Zero. Go ahead, what's the problem?
	CO-PILOT:	*Did you call? Clipper One Six Zero.
0909h16	MONTREAL:	*Oh, disregard, Clipper.

Once they had received their clearance to return, their attention was focused on the now worsening conditions in the cockpit:

0909h19   PILOT: It's still getting thicker. Isn't it?
ENGINEER: Seems like there could be equipment on fire.
PILOT: There is no smoke in these smoke detectors though . . . Is there?
0909h29   ENGINEER: Yes. There is now.
PILOT: There is?
ENGINEER: Yeah.
PILOT: Where would that pick it up from? Back there, or . . .
ENGINEER: Well, it's probably going up this way, and coming back around . . .
CO-PILOT: Yeah.
PILOT: Yeah.
ENGINEER: . . . through the forward one.
0909h45   ENGINEER: Turn the equipment cooling blower off. I think you don't need to go in the . . . back then.
PILOT: Right.
ENGINEER: Because it should pop a [circuit] breaker some place.
PILOT: Yeah.
0909h58   ENGINEER: We oughta go on oxygen [put the masks on]. This fucker's getting a little thick, eh?
CO-PILOT: I think so too.
PILOT: Just wait 'till we . . . Go ahead.
0910h04   PILOT: *Pan Ops from the Clipper One Sixty.
OPERATIONS: *One Sixty. Pan Ops New York. Go ahead.
PILOT: *Yes, sir. We just got our clearance to . . . ah . . . for a one eighty [degree turn]. We're coming back to New York and it seems to be getting a little thicker in here.
OPERATIONS: *New York. Clipper One Sixty. Understand that you're turning around now and returning to New York, and the smoke

	is thicker. Ah . . . will you be requesting the [crash] equipment on arrival?
	CO-PILOT: If we can't increase our ventilation.
	PILOT: *Ah . . . we'll let you know a little later on [about the crash equipment]. I think we have a few minutes. We're just up around Sherbrooke . . . between Sherbrooke and Cambridge right now, so we'll have another twenty minutes or half-hour.
	OPERATIONS: *Very good, sir. Thank you.
0910h50	PILOT: Jesus Christ. It *is* getting heavy.
	CO-PILOT: Huh?
	ENGINEER: I think we'd better take it to Boston.
	PILOT: Yeah.
0910h58	ENGINEER: This fucking thing is thick back here.
0911h00	CO-PILOT: *New York. This is Clipper One Sixty.
	PILOT: And tell 'em we wanna get down [to a lower altitude] and head for Boston.
	OPERATIONS: *OK. Go ahead.
	CO-PILOT: *Yes, sir. I think we're gonna take this thing into Boston. The smoke is getting too thick.
	OPERATIONS: *Understand. You're gone to Boston 'cause the smoke is too . . . Stand by one.
0911h17	CO-PILOT: *Boston Center. Clipper One . . .

At this point the co-pilot abruptly stops talking. A rustling noise on the tape indicates that he was donning his oxygen equipment. The sounds of breathing inside the mask become clearly discernible.

The engineer by now had also put on his mask. The smoke in the rear section of the cockpit had suddenly worsened:

0911h33 CO-PILOT: *Boston Center. Clipper One Sixty

		requesting direct Boston and . . . ah . . . requesting descent.
0911h40	BOSTON:	*Clipper One Sixty. Ah . . . Roger. Stand by just one sec. And . . . ah . . . wha . . . How low would you like to go?
0911h46	CO-PILOT:	*Ah . . . Say again please.
	BOSTON:	*One Sixty. Boston. Are you in an emergency or anything?
0911h54	CO-PILOT:	*Boston. *Please* give me a heading direct Boston at this time.
	BOSTON:	*One Sixty. Pick up a heading of . . . ah . . . one seven zero [degrees]. And when able, proceed direct Boston.
	CO-PILOT:	*Thank you very much.
0912h07	CO-PILOT:	*Ah . . . We'd like to start our descent also, if possible.
	BOSTON:	*One Sixty. Descend and maintain altitude of one eight zero [18,000 feet]. Correction. One nine zero.
0912h14	OPERATIONS:	*Sixty. Pan Ops.
	PILOT:	*Yessir. We're out of three one zero [31,000 feet] for one nine zero [19,000 feet].

The density of the fumes was increasing. Smoke was seeping into their masks. The usual headset microphones were thought to be causing more problems than they were worth. The crew decided to change to an inter-cockpit telephone system, hoping this would keep the smoke from getting into the breathing equipment. While the pilot and co-pilot were organizing themselves, the engineer answered a call from the Operations Room:

0912h20	ENGINEER:	*Pan Ops. Go ahead.
	OPERATIONS:	*Are you requesting the [emergency] equipment on arrival at Boston, sir?

He did not know the answer, and the pilot was in the process of adjusting his gear:

0912h25 PILOT: Do you guys want to get your goggles?
0912h33 ENGINEER: Do you want the [emergency] equipment on arrival at Boston? Probably wouldn't hurt, huh?
PILOT: Stand by one. I don't . . . know. What did . . . How's the smoke doing?
0912h43 ENGINEER: The son-of-a-bitch is full back here.
0912h48 PILOT: Better have the equipment.
0912h52 ENGINEER: *OK, Pan Ops. We want the equipment at Boston. Ah . . . the cockpit's full back here.
OPERATIONS: *OK. We're on the phone to them now.

The next three minutes or so were a struggle. With the badly designed and ill-fitting masks—all that the F.A.A. had certified and Pan Am had supplied for such emergencies—they coughed, choked and spluttered their way through the descent check list as the Boeing 707 continued to lose height. While the equipment was being readied and the navigation radio receivers reset, the conversations spasmodically returned to the dreaded subject:

0913h49 ENGINEER: OK. Fire warning. I'm gonna check the fire warning.
PILOT: Go ahead.

The sound of the bells reverberated loudly around the cockpit.

ENGINEER: OK.
ENGINEER: Test the instrument warnings.
0913h53 BOSTON: *One Sixty. Ah . . . Boston. Roger. Can you say again the . . . ah . . . nature of your emergency?
CO-PILOT: *Ah . . . We have smoke in the cockpit at this time.
0914h01 BOSTON: *Sixty. Roger.

0914h25 CO-PILOT: *Boston Center. Clipper One Sixty.
BOSTON: *One Sixty. Boston Center. Ident.
CO-PILOT: *Identing. And . . . ah . . . please . . . ah . . . just keep me on this frequency. It's too hard to change.
BOSTON: *OK. I'll keep you on this frequency. Roger, sir. Fly direct Kennebunk. [The air route is] Victor one thirty-nine skipper. Boston.

0914h42 CO-PILOT: *Kennebunk. Ah . . . Victor one thirty-nine skipper. Boston. Roger.
BOSTON: *And . . . ah . . . understand you have smoke in the cockpit, sir?
CO-PILOT: *Affirmative.

0914h52 BOSTON: *Maintain one nine zero [19,000 feet altitude]. Report reaching.
CO-PILOT: *Roger.

0916h26 PILOT: How does it look in the back, Dave?
ENGINEER: It's full.
PILOT: Is your D.M.E. [Distance Measuring Equipment tuned] in on Boston?
CO-PILOT: No, it's not coming in.
PILOT: OK. Watch the airplane. I'm gonna get my Boston plates [his landing charts].
CO-PILOT: You bet.

0916h56 PILOT: Smoke detector showing much?
0916h58 ENGINEER: No. Ah . . . it's showing the same as it was.
ENGINEER: We're somehow gettin' it up through the floor from down below, and it's goin' in the back [into the main cabin] . . . I think.

0918h34 BOSTON: *One Sixty. Ah . . . what is your . . . ah . . . altitude now, please. And if I can be of any assistance in any manner, let me know.

0918h47 CO-PILOT: *Ah . . . we're at one nine zero, and it's fine for us.

	BOSTON:	*Real fine. OK. Thank you.
0919h01	PILOT:	*We'd like to get down as soon as possible so we can burn off some fuel.

          BOSTON: *One Sixty. Go ahead.

          PILOT: *Yes, sir. We'd like to get down as soon as possible so we can burn off some fuel rather than dump.

          BOSTON: *I'm co-ordinatin' with the . . . ah . . . the lower [air space controllers] sector now.

          BOSTON: *Clipper One Sixty. Descend and maintain one zero zero [10,000 feet altitude].

0919h29   CO-PILOT: *Down to ten thousand. Clipper One Sixty.

The engineer had been continually searching around the dials and fuses on his control panels for a sign of trouble with the electrical or electronic equipment:

0919h45   ENGINEER: I can't find a thing wrong back here.

          PILOT: What's that?

          ENGINEER: *I can't find anything wrong.*

          PILOT: OK. Ah . . . maybe it's a package [of cargo?]?

          ENGINEER: Could be.

          PILOT: Ah . . . you didn't get in to open the door into the back section [the area which was laden with freight] did you?

          PILOT: Ah . . . they're supposed to be flame resistant, or fire resistant, anyhow. Aren't they?

0920h06   ENGINEER: Well, I . . . I looked back there. The smoke . . . there's more smoke back there now, but there's none up here.

          ENGINEER: It must . . . It's in the lower forty-one some place.

          PILOT: I think so.

0921h11	BOSTON:	*One Sixty. You anticipate flying . . . ah . . . locally to burn off fuel?
	PILOT:	*Ah . . . negative. We . . . Negative. We're coming right in.
	BOSTON:	*Yes, sir.
	PILOT:	*Ah . . . we would like as low [an altitude] as possible to burn it off as we're coming down and in.
0921h42	BOSTON:	*Sixty. Descend and maintain six thousand [6,000 feet altitude].
0921h45	CO-PILOT:	*Down to six thousand. Clipper One Sixty.
0921h52	ENGINEER:	Maybe we should advise the fire department that we suspect electrical fires?
0921h58	BOSTON:	*Sixty. The . . . ah . . . Boston . . . ah . . . weather is 4,000 [feet] . . . ah . . . scattered [clouds]; visibility 15 [miles] plus; runway 27 is the active runway, 33 left is available; the winds, two eight zero [degrees], ah . . . stand by the winds. Ah . . . two eight zero variable to three one zero [degrees], ah . . . at 15 with gusts to two five [knots]; altimeter setting, two nine seven . . . ah . . . four [inches of mercury]. The current altimeter is now two nine seven one.
0922h21	PILOT:	How long is [runway] 31? And how long is two seven?
	CO-PILOT:	Twenty-seven is 7,050 [feet], and 33 is 10,000.
	PILOT:	We'll take 31 . . . ah . . . 33. Runway 33.
0923h40	ENGINEER:	*Ah . . . Pan Ops. Clipper One Sixty.
	OPERATIONS:	*Clipper One Sixty. Go ahead, sir.
0923h48	ENGINEER:	*OK. We suspect this problem is electrical and it's in the forward end of the airplane. It's either the lower forty-one or the forward cargo hold, it seems like. There's

quite a bit of smoke in the cockp . . . the . . . ah . . . cabin . . . but . . . ah . . . there doesn't . . . there isn't too much in the cockpit right now.

OPERATIONS: *Ah . . . Roger. Roger. I have the equipment standing by. And what is your ETA, sir?

ENGINEER: *About thirty-five [0935h]. And have 'em open the lower forty-one when we get there, and . . . ah . . . stairs up to the front door. It doesn't seem to be that much of a problem.

0924h22 CO-PILOT: *Clipper One Sixty is requesting 4,000 [feet altitude].

BOSTON: *Sixty. Understand. Four thousand. We're trying to clear it with Pease Approach [controllers] now, and . . . ah . . . descend and maintain 4,000.

Because they had made the request to continue at lower than normal airways limits, it became the responsibility of the flight crew to watch out for any private low-altitude planes that might happen to be around.

The engineer started the approach checks. And soon, with the equipment prepared, the Boston controller was again asking whether they would care for a still lower altitude:

BOSTON: *Sixty. Two thousand is available. Just let me know.

0925h30 CO-PILOT: *Clipper One Sixty is out of 4,000 for 2,000.

BOSTON: *Roger.

PILOT: I'd rather bump a little and get down there and burn some of this fuel off.

CO-PILOT: Say again the landing gross weight?

ENGINEER: OK. It was 280, but we're not burn-

ing it [the fuel] up very fast.
ENGINEER: Call it two seven five [275,000 pounds] for landing.
CO-PILOT: OK. That looks like 170 [miles per hour] for the landing [speed].
PILOT: Ah . . . throw out the gear please. [Lower the landing wheels.]
CO-PILOT: Gear coming dow . . .
PILOT: . . . *Hold it. Hold it.* I'm sorry. Wait 'till I slow it [the aircraft] down. We'll tear the fucking doors off [the landing gear wheel wells].

0927h59 PILOT: I don't smell that smoke as much now. There doesn't seem to be as much, does there?
ENGINEER: Ah . . . ah . . . it doesn't seem to be as much.
PILOT: Huh?
ENGINEER: *It doesn't seem to be as much.*
0928h34 ENGINEER: OK. The engineer's check is complete; the approach check is complete; the landing [check] is next.

The slight improvement in the conditions inside the cockpit was a blessing, but not for long. The Boeing 707 was now on a course above the Atlantic Ocean near Hampton, New Hampshire, heading directly toward Logan Airport. The airfield's emergency equipment was being readied for the arrival of the freight flight, the crews checking to make sure that they had electrical fire-fighting foam on board their trucks.

0930h36 PILOT: All of a sudden it is getting worse in here.
ENGINEER: Yeah.
ENGINEER: It's somewhere down in the lower forty-one.

0930h46   ENGINEER: Tell ya what. Turn the radar off . . . the doppler's [radar] off . . . anything you don't need. Let's shut 'em down.
0931h20   APPROACH: *One Sixty. Boston Approach Control. Radar contact 35 [miles] northeast of Boston. Proceed direct Boston. Maintain [altitude] 2,000. Are you declaring an emergency?
0931h29   CO-PILOT: *Negative on the emergency. And . . . ah . . . may we have runway three three left?
0931h33   APPROACH: *That is correct. You can plan three three left. Understand negative emergency. Maintain 2,000 [feet altitude] and . . . ah . . . expect a visual approach to runway three three left. The Boston altimeter is two nine seven one [inches of mercury]; the wind is two niner zero [degrees] at one eight [knots]; the Boston weather, 4,000 [feet altitude for] scattered [clouds]; visibility more than one five [miles].
0931h50   CO-PILOT: *Roger. Roger, Boston. Clipper One Sixty.

0933h42   ENGINEER: It doesn't seem to be getting any worse.
0933h44   PILOT: No. No, but I don't think it is getting any better. Is it?
0933h46   ENGINEER: No. It's not getting any better.

0934h20   APPROACH: *Clipper One Sixty. What do you show for a compass heading right now?
0934h23   CO-PILOT: *Compass heading at this time is two zero five [degrees].
0934h26   APPROACH: *OK. Fine. And will you accept a vector for a visual approach to a five [miles] final? Ah . . . will that be sat . . . compatible with you?

Suddenly, in the middle of the approach controller's words, a tremendous screeching noise started up in the crew's headsets. It seemed to be coming from a test circuit built into the cockpit's electronic equipment. The sound was loud and piercing but, with full concentration, it was just possible to make out what was being said:

0934h31   PILOT: I didn't hear that. Try him again.
0934h33   CO-PILOT: It's the first time we lost that circuit.
0934h35   CO-PILOT: *What was that, Approach?
0934h36   APPROACH: *Will you accept a vector for a visual approach to a five final for runway three three left? Or do you want to be extended out further [than five miles to turn in towards the runway]?

At this point, the cockpit voice recorder abruptly stops. However, the radio conversations were captured on sound tape back at the airport.

0934h43   PILOT: *Ah . . . negative. We want to get in as soon as possible.
0934h46   APPROACH: *OK. Proceed to Boston VOR [a navigation radio beacon]. Advise when you have the airport in sight. And Clipper One Sixty, you're number one [the next aircraft] for runway three three left.
0934h53   CO-PILOT: *Roger. Clipper One Sixty.
0934h55   APPROACH: *Are you able to maintain 2,000 [feet altitude]?
0934h57   CO-PILOT: *That's affirmative.
0934h58   APPROACH: *OK. Fine. There will be traffic at ten o'clock, one zero [miles] westbound. An Air Canada Viscount, descending to 3,000 [feet].
0935h05   CO-PILOT: *Roger.

0935h46 APPROACH: *Clipper One Sixty. Advise any time
y u have the airport in sight.

At 0939h, Captain Bruce C. Buck was taxiing his Eastern Air Lines Boeing 707 out from the Boston terminal building toward the end of Runway 27 in preparation for a flight to Newark, New Jersey:

"We were then instructed to 'hold' on taxiway Charlie, well clear of the runway. I stopped the aircraft practically where we were.

"I looked to the east and noticed what appeared to be a Pan American 707-type airliner. It was making a slight turn to apparently align with Runway 27. I estimated the altitude to be about 2,500 feet, at a distance of 10 miles from the airport. I saw no signs of smoke coming from it at this time. It appeared to me that it was flat, slightly nose-down, and moving very fast. As it descended in a flat attitude, it began to 'Dutch-roll.' And as this intensified, yawing [nosing from side to side] increased and it then changed to a nose-high attitude.

"At two miles out on the final approach, it appeared he was experiencing real control problems. The machine was 'Dutch-rolling' badly, and the nose was quite high. The speed seemed fast and the aircraft had too much altitude for a Runway 27 landing.

"At one mile out and at about 500 feet altitude, it looked as though a great deal of power was added because large quantities of black smoke came from all four engines. The aircraft leveled off, the nose lowered to a flat attitude and it looked as though he was going to attempt a 'go-around.' He was very high for a Runway 27 landing at this point.

"As they neared the threshold it was 'Dutch-rolling' badly and the nose was going higher and higher. The aircraft rolled to its right, then sharply left—in excess of ninety degrees. The nose came up. The machine appeared to stall.

Suddenly, the nose dropped sharply. The left wing and nose section struck the ground at the same time, only feet short of the runway. The plane was nearly vertical at impact. It immediately exploded and was engulfed in flames.

"As the whole episode was such a ghastly sight to watch, and as passengers in my aircraft could not possibly have missed seeing it, we returned to the ramp area, parked at the gate we had recently left, and I think everyone was glad to get out."

Thomas H. Young worked for the U.S. Coast Guard at its Boston Bay headquarters. He had, three hours before, taken over from the night-duty watchman in the Operations Room. Their log was very sparse—mainly routine messages from the various vessels, until, at 0940h, the radios started buzzing:

0940h   GB TI PA   Something seems to have blown up at Logan.

0942h   GB TI PA   Have a report on FM radio Salem Control that a plane went down in the vicinity of Snake Island.

0943h   GB TI PA   Have two boats under way now.

0948h   GB TI   Coast Guard auxiliary vessel Tiger Shark reports heavy black smoke coming from the end of Logan's runways. Proceeding there now.

0955h   GB TI   Auxiliary vessel Tiger Shark—seems to be a cargo plane. Have no sightings of people in the water. Plane seems to be mostly on land. Some debris in water, but have no sightings of people.

0958h   GB TI 40533   On scene at this time.

0958h   GB TI TIGER SHARK   Notify fire department that the pier at the end of the runway is on fire.

1013h   GB TI 31023   Have gone as far as we can go. Quite a bit of debris in water and up close to land, but cannot get any closer.

1013h	TI GB R	Stand by to keep sightseers out of the way and general policing of the area.
1015h	GB TI	Port Authority requests to know if you are ordering a helo due to the fact that there are deposits of debris in close to shore and you cannot get in all the way.
1015h	OP NOTE:	OOD notifying Port Authority that helo is not necessary.
1020h	40538 TI GB	OOD advises that you take charge as on-scene commander to keep people out of the area and take any cargo found on board.
1025h	GB TI 31023	Have recovered one body. Bringing it to the fire department.
1030h	31023 TI GB	Where exactly did you find the body?
1030h	GB TI 31023	Off the runway near the approach lights. Do you know the name of the runway? It was near the approach lights.
1035h	GB TI 31022	We have picked up three sacks of mail.
1035h	GB TI 40538	We have picked up some plastic bottles with some kind of chemical in them.
1035h	TI GB	All salvageable stuff you pick up, we want you to write down exactly where you picked it up. Hold on the boat and we will eventually bring you back to the base and distribute the items to the proper authorities.
1055h	GB TI 40538	These bottles we found were nitric acid.
1055h	TI GB R	Handle with extreme care and keep well away from all electrical equipment.
1102h	GB TI 40538	Advise OOD that debris seems to be going up on land at Port Shirley and would be advisable to get some personnel over there to maintain observation.
1125h	40538 TI GB	OOD advises that you can release the 40567 and also do you have a better description of those nitric acid bottles?
1125h	GB TI 40538	They are seven-pint glass bottles, red

and white and a brown one that seems to be about the same net weight of approximately six and a half pounds.

At 1127h, the patrol boats reported that the seas in the Bay were becoming very choppy. All but one were then recalled to headquarters.

As the bits and pieces picked up during their search were landed, a formidable list of items accrued, including one eighteen- by thirteen-inch piece of metal and six small cases of B & A nitric acid, code number 108-2677 with six and a half pounds in each.

The nitric acid, an extremely corrosive chemical, apparently had somehow fallen out of the aircraft during the minutes preceding the crash. As soon as the investigators from the National Transportation Safety Board heard about this, they immediately impounded the flight's cargo manifests:

Number of Packages	*Nature of Goods*
60	Plasticatros
5	Electronic computer parts
17	Electronic components
274	Chemicals
1	I. B. M. parts
1	Electronic computer parts
1	Electronic computer
20	I. B. M. parts
1	Machine
7	Electrical equipment
1	Military stores

Because of the evident hazards involved, both national and international airways' standards stipulate that materials mentioned on the *Restricted Articles* list must be separately notated on a report called *Notification for Loading Restricted Articles*. And this must be signed by the captain of the aircraft on which they are to be carried so that he and his flight crew are well aware of the potential dangers.

The partly charred remains of all but one copy of the *Restricted Articles* documentation (unsigned) was found in Clipper 160's burned-out wreckage. Captain Zammett and his officers were completely unaware of what they had on board.

The *Restricted Articles* list read as follows:

Product	Number	Quantity	Hazard Label
Butyl acetate normal	1 piece	70 pts. tot.	Comb. L. label
Poison liquids nos.	1 piece	56 pts. tot.	Pois. B. label
Isopropanol	10 pieces	56 pts. each	Red label
Hydrogen peroxide	16 pieces	7 pts. each	White label
Acetone	10 pieces	70 pts. each	Red label
Nitric acid	160 pieces	4½ pts. each	White label
Methanol	3 pieces	70 pts. each	Red label
Acetic glacial acid	9 pieces	70 pts. each	White label
Hydrofluoric acid	50 pieces	4½ pts. each	White label
Xylene	4 pieces	70 pts. each	Red label
Sulphuric acid	60 pieces	4½ pts. each	White label

With all that potentially hazardous stuff in the aircraft, anything could have happened. But of all of it, the nitric acid was the most interesting—and the most dangerous. Bottles of it—some broken, others that had lost their labels in the water—seemed to play a major part in the disaster. Rudolph Kapustin, the investigator-in-charge, decided to do a little "historical research."

In October, 1973, the National Semiconductor Corp. of Scotland had cabled a long list of materials it required to its parent company in Santa Clara, California. National Semiconductor of California had a standing agreement with Allied Chemical Corporation for the supply of such requisites. So it ordered 160 bottles of sulphuric acid, 32 bottles of A-20 stripping solution, and 525 pints of T.C.E. from Allied's Los Angeles factory. At the same time, National Semiconductor asked for a variety of different products, including some 40 cartons of nitric acid in glass containers, from its Marcus Hook, Pennsylvania, plant.

Allied did not do much aerial shipping—the special packaging that was required would have interfered with its production lines. It therefore contacted Lyon Commercial & Export Packing, which agreed to rebottle and prepare the acids so they would meet air freight regulations.

Meanwhile, National Semiconductor phoned InterAmerican Forwarding Corporation to make arrangements for the Los Angeles order to be shipped to New York, where it was to be added to the Pennsylvania materials as combined transatlantic air cargo.

The bottles of sulphuric acid were crated by Lyon exactly as they arrived from Allied Chemicals—in wooden containers, with an absorbent material called vermiculite filling the spaces and acting as an internal cushioning. The wrong type of "Corrosive Liquid" labels were then stuck on the outside. Box numbers, arrows denoting "This Way Up" (but without accompanying labels), the airwaybill number, and the address to which they were eventually to be delivered were all painted on them. The boxes were secured to fork lift skids with nylon tape. InterAmerican had instructed Lyon to take the shipment to Burlington Northern Air Freight.

Burlington, once the consignment had arrived, glued Pan Am "lot labels" on the exterior of each box before loading the skids into an airways palletized container, sealing it, writing out an airwaybill describing the goods as "electrical machinery," then transporting it all to the T.W.A. freight dock at Los Angeles International Airport—initial destination, New York.

Over on the East Coast, InterAmerican Forwarding Corporation had cabled a New York company, the Seven Santini Brothers, that a truck would be delivering a load of chemicals from Allied's Pennsylvania factory. Santini agreed that, when it arrived, they would overpack the shipment in accordance with air freight regulations.

Since the office manager at Santini's had no noncombustible cushioning substances as the rules required, he presumed that sawdust would be all right on the basis that "if it was OK for red label materials [inflammables], it was OK for white labels [corrosives]."

He then realized they had no metal cans in which, the regulations stated, the acid bottles must be sealed. So he telephoned the sales and service manager at Lyons in Los Angeles and was told that tins were not really necessary.

The nitric acid, together with the remainder of the order, was duly packed and marked—without "This Way Up," "For Cargo Aircraft Only," and "Corrosive Liquid" labels on the outsides—and delivered to Pan American's Cargo Center at Kennedy Airport. Pan Am personnel unloaded the crates and signed for them. They were expecting two lots of chemicals—one from Santini, the other from Los Angeles—now identified to T.W.A. as "electrical appliances" following a late phone call from Burlington. By 1930h that Friday, both consignments had arrived. They were scheduled out on the following morning's Flight 160 for Europe.

This was an unusually large amount of *Restricted Articles* for the Pan Am Cargo Center to handle. On looking it over, those in charge decided that it could best be arrayed on three airway pallets.

The loads had to be so constructed that they would fit around the concave curvature of the plane's inside fuselage. This meant that, even though both deliveries had come neatly bundled on fork lift skids, they would have to be taken apart and repacked on Pan Am pallets, remembering, of course, the problems that had to be solved so that as much as possible could be jammed into the aircraft's available space.

As the investigation reveals:

"At some point during the repacking, it was discovered that the boxes along the outer edges of the pallets were stacked too high to fit within the contour of the fuselage. Four of the loaders stated that they were informed by their supervisor to lay the boxes on their sides. It was estimated that eight to ten boxes were laid side-on on each of two or three different pallets [and possibly including a fourth they found necessary because they underestimated the size of the load]. One of the packers said he was assigned, during the following [midnight] shift, to wrap three pallets in protective plastic sheets which had not been completed by the prior shift. He said that he informed his supervisor that the boxes were loaded 'against the arrows,' but he was told to wrap them as they were."

At this point, the investigators were still not sure what had happened to cause the disaster. They realized that the airways' *Restricted Articles Regulations* had been insouciantly ignored.

Since a few of the nitric acid bottles, still in their original containers, had been recovered from Boston Bay, they considered drying these out and conducting a test. The notion came to them after they had revisited Pan Am's cargo terminal at Kennedy Airport. While being shown around by one of the heads of the department, they noticed that some crates with clearly marked "This Way Up" arrows were lying on their sides. When they discovered that a rescued nitric acid bottle had a loose cap, they decided to test their theory.

After drying the crate and sawdust and imperceptibly loosening the plastic stopper on the bottle only as far as the metal seal-band would allow, they repacked it as it had been delivered to Pan Am by Santini.

On a clear patch of ground near the Delta Air Lines hangar at Boston airport, they tipped the box upside down and timed the results.

After seven minutes, a bluish-white smoke was observed coming from around the lower surface of the container.

After eleven minutes, the smoke downwind had a smell similar to that of burning wood.

After thirteen minutes, the white smoke flared profusely, changing momentarily to orange in the vicinity of the box.

After fifteen minutes, the quantity of smoke subsided.

After seventeen minutes, the smell of burning wood returned and a large pool of acid started spreading around the area nearby.

After nineteen and a half minutes, flames were visible from near the bottom of the box.

And after twenty-one and a half minutes, sheets of fire burst through the top.

The investigators returned to the hangar where all the collected wreckage had been stored. Among the debris was the badly corroded remains of a pallet, retrieved from the Bay by the Coast Guard patrols. Just visible on it was the identification number "013." Pan Am's records showed that it had been assigned to position number six on Flight 160. This was the one on which a large part of the nitric acid had been stored—the one on which the boxes that would not fit properly had been tipped on their sides.

The remainder of the puzzle fell into place.

Pallet number six had been positioned directly above the central wing area of the Boeing 707. Further investigation showed that as the nitric acid spilled from one or more side-on containers, probably as the air pressure inside the cabin decreased during the first part of the flight, it ignited the sawdust and possibly the packing crates, then, running along the cabin floor, burned its way through the metal and dripped into the cargo holds below.

During the last minute in the air, huge sections of the machine's underfuselage had been eaten away. The pallets of freight in the main cabin had collapsed down into the lower compartments, the floor disintegrating as a result of nitric acid corrosion. As the holes in the underside, assisted by the fires from the burning sawdust and wood, enlarged and increased in number, first the bags of mail and then whole pallets, complete with their contents, dropped from Flight 160 into the waters of Boston Bay. In those last few seconds, nearly half of the entire contents of the airplane had fallen out. No wonder the pilots had lost control.

Now, just eighty-eight days later in Auckland, preparing for the trip to Pago Pago, Clipper 806 was also loaded with

Restricted Articles. And like John Zammett before him, "Pete" Petersen departed the ramp completely unaware of what was really in the freight bays.

3

They were fourteen minutes behind schedule as acting copilot Phillips lifted the 707's nose gently but decisively into the air.

The hop to Pago Pago should take three hours and fifty minutes. Once they were airborne and some altitude separated them from terra firma, the "No Smoking" and "Fasten Seat Belts" signs were extinguished and the stewardesses began taking orders for pre-dinner drinks.

Roger Cann struck up a conversation with the man on his right in 16-C. He was an American by the name of Charles Culbertson; young, just under six feet tall, with short hair and glasses. Culbertson told the Canns that he was returning to Honolulu after three months at sea—he was an oceanographer. But following their introductory niceties he seemed to have very little to talk about so, while the refreshments were being served, Roger Cann took in the activities elsewhere.

Directly in front of them, in 17-A and 17-B, was an elderly couple—Mr. and Mrs. M. P. Merrill—though the Canns had no conversation with them during the journey and did not even learn their names until later.

In the row behind—in 15-A, 15-B and 15-C, and across the aisle in seats 15-D, 15-E, and 15-F—were, evidently, six friends. They were in the fifty-to-sixty age bracket. Their comments centered on the city of Los Angeles. Though he could not clearly make out what they were saying, Cann surmised from the accents that one of them was French.

Forward of the Canns, at about row 18 and on the right side of the cabin, was an animated group of young people. They were speaking loudly and there was a lot of giggling and laughing. Cann decided they were New Zealanders because he noticed that they paid for their pre-dinner drinks with money from that country.

In front of them was Dick Smith in seat 20-F, next to an emergency exit window. Position E beside him was vacant, but in 20-D there was a young woman from Australia and Smith briefly talked with her. Across the aisle in 20-C, 20-B, and 20-A, two Samoan girls, about eighteen to twenty years of age, were amicably chatting with a twenty-five-year-old Japanese man.

Soon Smith decided to go to the lavatory. He strolled slowly to the rear of the cabin, looking about him as he went. He did not know the Canns, so he passed them by with hardly a glance. A few rows behind them, however, he recognized the Lewis family, whom he had met in Pan Am's Clipper Lounge at the airport. They had been stationed for the last four-or-so years at Alice Springs in central Australia, and now they and their three daughters—twenty-one, nineteen, and either five or six years old—were returning to America. They were planning to live somewhere in Texas.

After returning to his seat, Dick Smith studied the situation around him once more. Sitting in the aisle seat in row 21 was a portly businessman type. In the two window positions in front of Smith were the young American mountaineers, even now readying themselves for an after-dinner sleep. At about row 23, and spread either side of the passageway, was another family, consisting of mother, father, and three small children—the Ortons.

Soon the stewardesses arrived with Smith's before-dinner cocktail. While sipping it he reviewed in his mind everything he knew about his stopover point.

The archipelago of Samoa is in the south-central Pacific, ten and a half flying hours from San Francisco, five and a quarter hours south-southwest of Hawaii, three and a half hours north-northeast of Auckland, and three hours north of Tahiti.

Eastern (or American) Samoa is a cluster of seven small islands with a total area of seventy-five square miles. The main one is called Tutuila, and on its south-central coast is the capital, Pago Pago. The city is situated on one of the most picturesque natural steamer ports in the South Pacific, being overlooked by Mount Rainmaker, a 2,000-foot-high volcanic peak on the other side of the bay. This was the setting for Somerset Maugham's short story, *Miss Thompson,* later turned into a film called *Rain.* And *that* it certainly does in this tropical region. December to April is the monsoon season. The drenchers, however, are usually quick in coming and equally rapid in their departure.

The average mean temperatures are ideal. All year around, the daily minimum varies by only one degree on either side of 76 Fahrenheit. Maximums range from 84 to 86.

Besides being the home of Robert Louis Stevenson and Margaret Mead, possibly the best-known features of the area are the wide, long, oh-so-white sandy beaches. Lining them on the one side are the deep azure blues and emerald greens of a polution-free ocean; on the other is the lush jungle with its shading, waving palm trees.

Unlike the Hawaiian Islands, ruined by thoughtless modern development, Samoa still maintains the folklore and outlook that have been part of its everyday life for the last two hundred years. The only modern architecture is in the towns. Traditionally built native villages dot the main beaches wherever the fishing reefs are not too far out from the shoreline. The settlements are sometimes difficult to spot, half-hidden from view among the palm trees and coconut groves. The houses look something like brown beehives, suspended above the regularly rain-soaked ground on stilted foundations of stone. Without electricity or telephones, they are a far cry from what most city dwellers would relish. Business suits are practically taboo. Almost everyone wears garish sports clothes or beachwear, even when working in the town center.

The beauty of the islands, though, is their perfumed forests, stretching from the very edge of the villages as far as the eye can see, spreading their lush greenness up the sides of the hills and mountains that form the backdrop. The visitor is in a world apart as he strolls along any of the few narrow roads that penetrate the thick foliage. He must never be seen eating in the streets—that is the height of bad manners. And if he should be invited into one of the shanties for a cup of the ceremonial *kava*—made from the root of the Ava tree—tradition states that before drinking he must first pour a small amount of it on the ground. Should there be any left when he has finished, that, too, must be returned from whence it originally came.

The local foods, naturally, use as their chief ingredients the islands' main produce: coconuts, bananas, taro. The three are combined into a dish called *pulusami*. Another, consisting of breadfruit, taro, bananas, fish and roast pork, is named *fia-fia*. And most gourmets have at least heard about, if they have not relished, *poi* (mashed ripe bananas mixed with coconut milk).

"The purchase of liquor is limited to those who have permits issued by the Government," says the Pan Am Travel Guide ominously. But then everything, including the bars, is closed on Sundays, the complete day of rest.

The most exciting sights are visible from the aerial cable car that starts at Solo Hill, at the edge of Pago Pago, and climbs slowly all the way to the top of 1,600-foot-high Mount Alava. From the pinacle, the entire area known as American Samoa can be viewed, surrounded by the blue waters of the Pacific that stretch in all directions as far as the eye can see.

A short distance from Tutuila, to the northwest, is the "other Samoa"—Western Samoa. It is a completely independent nation, interconnected with its neighbor not only by name, but also by its Polynesian ancestry, its traditions, and by boat and plane. Western Samoa is, in the main, two large islands: Savai'i and Upolu. It was on the latter, at Vailima, that Robert Louis Stevenson lived. He is buried on nearby Mount Vaea.

Between them, the two Samoas have a population of about 178,000; of this figure, just 28,000 reside in the American territory.

For Dick Smith, the stopover was going to be too short. Roger and Heather Cann at least would have a full week in which to take in as much as they possibly could.

But now, in the 707's cabin, the stewardesses were preparing to serve the meal.

Roger Cann walked back to the lavatory. As he passed the galley, he noticed that the "cook," an attractive blonde girl, was busy at work, assisted by a Japanese stewardess.

On his return, he happened to glance a row or two in front of where he had been sitting. There was no seat at 18-A, but instead a space, next to the rear of the over-wing emergency windows on his side of the fuselage. Three kit bags filled the area, blocking easy access to the exit. One of the haversacks contained a squash racket or a similar piece of sports equipment—its handle was jutting out. All in all, there was enough luggage in that narrow clearance to make Cann say to himself *"That shouldn't be there."*

Up in the cockpit, meanwhile, a red warning light was flashing on and off. It was a door-open indicator telling the crew that the rear main passenger entrance was not fully or properly locked. Engineer Gerry Green was sent back to inspect it. But after he had a thorough look at the air-seals, the handle's position and the panel's placement, he reported back that it appeared to be closed securely; perhaps it was simply a malfunction in the circuitry itself. After checking to make sure the pressurization in the cabin was holding, the crew marked in the log that there was a fault in the door-warning wiring.

2100h. The flight was continuing normally. The evening meal was finally served. The co-pilot, at the plane's controls, had been informed by radio that there was some slight turbulence up ahead and that they should expect one or two severe jolts. It was nothing that would cause great concern, he was advised.

After dinner, while the hostesses were collecting the trays, the passengers felt a bump. For safety's sake, the captain instructed the engineer to switch on the "Fasten Seat Belt" signs.

Once everything was cleared away, the main cabin lights were dimmed. Some of the passengers immediately settled in to catch a couple of hours' sleep before the expected landing in American Samoa.

Dick Smith is a rather nervous, fidgety person when confined to a small area for long periods of time. He was not nervous about flying—he had traveled many hundreds of thousands of miles in airplane seats; it was the restrictions in the space in which one could move that made him restless.

During the period after the meal he struck up a conversation with the Australian girl in 20-D. She was about twenty years old, plain looking, but with facial features that really came to life when she smiled or laughed. She was continually searching through the many parcels and packets she had brought with her, now stowed under her seat, to find some reading matter with which to while away the time. She was casually dressed in a skirt and blouse, and had a dark sweater draped across her shoulders.

Smith asked her where she was heading. She was an adventurer, she told him. She wanted to work her way around the world, visiting as much of it as possible, before settling down. She had already begun with a vengeance. Over the last few months she had seen a lot of New Zealand, employed as a maid in several provincial hotels, then as a waitress in an Auckland restaurant. She was now moving on once more. Her next stop was Honolulu, where she hoped to find something to keep her occupied for a while before continuing on to the United States mainland.

Smith brought up his pet subject, but she had little interest in it. Her only sports experiences had been equestrian. She had spent some of her school holidays with friends who owned a farm, she said. There, they had done a certain amount of horseback riding.

One of the hostesses, checking to see that everyone was comfortable, happened to overhear the conversation and told the girl Smith was an Olympics diving coach. Diving, the stewardess said, had been one of her favorite activities when she was at college.

The Australian then began talking to the Samoan lasses and the Japanese man sitting directly across the central aisle. The two Polynesians had been at school in New Zealand and were very excited about returning home. The one in seat 20-C was, as Smith later remembered, rather plump but very cheerful. Her friend in 20-B was smaller with finer features. Both were pleasant looking. They chatted to the Japanese chap in the window seat next to them, about their hobbies and their experiences in New Zealand and Samoa.

The Japanese gentleman was short but well built, with broad shoulders; like most orientals of his age he was neatly dressed and immaculately groomed. He told the girls in excellent English that he was on his way for a holiday in America.

Now that the turbulence had ended and the "Fasten Seat Belts" signs had gone out, Smith decided to do some writing. He wanted to let two of his closest friends know about his diving team's successes in the British Commonwealth

Games. The first letter was to a Dr. Sammy Lee in Santa Ana, California, even though he was sure he would see him in Los Angeles before the correspondence arrived. The second was to an R. Jackson Smith in New York, telling him of the modern diving facilities in such a remote spot on the globe as Christchurch.

At 2200h, James L. Prendergast, the principal maintenance inspector with the Federal Aviation Administration in Honolulu, arrived at the Pago Pago International Airport. He was in a van driven by Jay McLean, a pilot with South Pacific Island Airways, one of the local air taxi companies. Also with them was Cary Wade, a mechanic working for the same airline. Prendergast had made it to the terminal a little earlier than he had expected because of McLean's offer of a lift. He was due to catch Clipper 806 for his return to Honolulu.

Prendergast, who had been connected with aeronautics since 1950, held a commercial pilot's license, a mechanic's certificate, and a senior parachutist's diploma. The two air taxi employees dropped him off at the main building, agreeing to join him later in the coffee shop. While he checked in his baggage at the Pan Am desk, the others drove around and parked in the airport's lot.

Back on the 707's flight deck at 2201h, Jim Phillips radioed the Nadi high-altitude traffic director to report that they had passed the "Bagpipe" air-lane crossways at 2200h, that they were flying at a height of 37,000 feet and that they estimated they would be at the "Tonga" route intersection at 2232h. After a pause, the controller recontacted them to question the elevation figure that had been given to him just two minutes earlier. The operating co-pilot apologized—it was his error—they were at 33,000 feet.

After finishing his letters, Smith decided to take a walk along the dimly lit cabin. At the front of the tourist-class compartment he discovered the little old lady he had briefly chatted with at the bank in the Auckland airport. She was American, very short and petite, sixty to sixty-five years old. She had been on holiday in New Zealand, and now was to visit a daughter living in Polynesia. Smith smiled and waved as he strolled by.

Most people up front seemed to be asleep. Smith sauntered back. He neared the rows where the young American mountaineers were stationed in the window seats. They, too, had nodded off. They appeared to be comfortable enough—both were wearing old, faded jeans, once-garish shirts, and elderly multi-patched pullovers. One of them had on a pair of sandals. Even slumbering they looked exhausted. They might well have rushed directly back from their hillside camping site at the last possible moment to catch the Clipper for home.

By comparison, Smith himself was sartorially elegant in his air travel clothes: he wore a light gray sports jacket, a pastel blue shirt, a turquoise bow tie, a pair of grizzled muslin trousers, and the normal—regimental—black cotton socks and black shoes.

Seated just in front of the mountaineers was the Samoan family, the Ortons. The mother and father were on the left side of the cabin. They were both short and plump. The three children ranged in age from about five to nine. They were across the aisle from their parents, on the right-hand side. They seemed to Smith to be remarkably quiet, playing with a game the hostesses had given them. Smith leaned over the one in the seat next to the passageway and asked how they were doing. Fine, they responded.

It was now 2234h. The 707 had arrived at the "Tonga" air-lane junction. Again Phillips radioed Nadi. This time he reported their estimated arrival at the "Ladyfish" intersection as 2322h. A couple of minutes later they felt another turbulent bump—once more the "Fasten Seat Belts" signs went on for a short while.

The lighted warnings forced Smith back to his place. It was a good time to do some of the work he had brought with him, he thought. He retrieved his briefcase from under the seat and took from it a copy of a book he had helped to write, *Inside Diving,* together with a script he was working on and a photoprint of the plans for a huge swimming and sports complex that was about to be contracted for construction.

The book had only recently been published. Smith had taken a copy of it with him to New Zealand so that he could read it thoroughly for possible mistakes. The scripts were for a new idea of his, a teach-yourself-diving course on sound cassettes. He needed to finish planning the details fairly soon so they could be produced. But the most urgent of the three was the blueprints.

Besides his activities as a diving coach, Smith was also employed as a consultant to a small firm called the Mitchell Energy and Development Corporation, who were in the process of building a sports center in the new town of Woodlands, Texas. Smith had been retained as the swimming facilities adviser. He had taken a copy of the plans with him to New Zealand in order to go over them. He had found an error in them. Now, as the aircraft pushed its way through the high-altitude atmosphere 33,000 feet above the Pacific, Smith again checked the blueprints to make sure the mistake he had discovered was genuine.

At the Pago Pago International Airport, it had taken James Prendergast about thirty minutes to check-in his luggage and obtain his boarding card for Clipper 806, scheduled to depart for Honolulu at 0030h. He then walked briskly up the stairs to the terminal's coffee shop, where the two men from South Pacific Island Airways were waiting. It was an agreeable evening. The temperature was a rather coolish 79 degrees Fahrenheit, and though it had not rained, some dark, ominous shower clouds were massing.

Because of the pleasant weather, the three selected a table near a window space to wait for the plane's arrival. In the meantime, they talked and drank coffee.

At 2240h, the Nadi controller again radioed the 707 asking if the pilots would like to move up to an altitude of 37,000 feet. He remembered that Phillips, in his earlier report, had mistakenly given this figure and wondered whether they might have preferred the higher flight level. The copilot replied that they were quite happy at 33,000.

Back in the passenger cabin, Roger Cann was reading. His wife, next to him, slept. Not having flown before, he felt that things had been very smooth. He had not expected it to be so trouble free. There had only been those two instances of turbulence when the signs had come on, and then he had simply tightened his belt as instructed. He had not been at all nervous.

Smith was again on his feet, this time walking toward the back of the plane. There were the Lewises. He half-perched on the aisle-seat armrest in order to have a chat.

The most outgoing character in the Lewis family was the younger of the three girls—blonde, about five or six years old, effervescent, vivacious (even though it was extremely late in the day for her), talkative, electrified at the prospects of seeing the States for the first time, ecstatic about her new school, her new town, her new friends.

Looking across the passageway, Smith could see that the two elder Lewis girls were fast asleep. And why not? After the excitement of "moving day," the usual frustrations of getting this far, and the lateness of the hour—plus the constant to-and-fro rocking of the fast-speeding Boeing 707—they had every right to be. It was a wonder, he thought, that he wasn't at least a little drowsy himself.

4

Boeing's main plant, prior to its building a huge new complex at Everett some twenty-five miles north of Seattle, was at Renton, about halfway between the city and the Seattle-Tacoma Airport to the south. When you drive north from the airport, your first glimpse is of the end of the Boeing Field runway, just to the left of the downtown highway. Beyond the strip and covering most of the area on the other side of it are a number of factory buildings, the dominant one being much taller than the others and clearly marked as the *Boeing Space Center.* Around it, the constructions are lower and flatter, but internally much bigger.

The principal workshop, with a floor space of about 220,000 square feet and a ceiling some sixty feet high, is normally a hive of activity. Overalled engineers rush here and there with formed pieces of metal. Boxes of rivets, screws, nuts, and bolts, together with construction plans and miles of electrical wiring, are to be found on most of the trestled tables dotted around the place, usually right alongside large printed signs indicating the airframe numbers of the under-construction work nearby. Other boards state the job in progress and that still to be done, together with the timetables.

Standing on an overhead inspection platform looking down on the activity below stretching off into the distance, one's head is filled with sound—the gentle clanging of work tools interspersed with loud, disruptive screechings as the bits on the business ends of large industrial drills eat through metal. Most of the sections, being placed together like a giant jigsaw puzzle, were actually manufactured away from this plant.

Dotted around the walls at various points are computer terminals. These, besides keeping tab on employees' attendances, also lovingly memorize the information on control cards fed into them at regular intervals as each new part of the jigsaw is bolted, riveted, screwed, locked, or glued into position. In this way, the time taken and the progress of the job at hand is religiously recorded for daily viewing sessions by senior management.

It was in such fashion that construction number 19376 slowly became a living entity, formed out of the millions of prefabricated bits and pieces created elsewhere.

She was a Boeing 707, a model 321B. She was the ninety-eighth 707 the workers at Renton had slaved, sweated, and toiled over for Pan American World Airways, and the four-hundred-sixty-eighth of her kind.

Of the U.S. big three, the Boeing Airplane Company was "odd-man-out." Both Douglas and Lockheed had begun their operations in or near Los Angeles. William Edward Boeing, on the other hand, was a Washingtonian by adoption.

It was not until July of 1914 that Boeing first tasted the thrill of flying. An "adventurer" had navigated a Curtiss seaplane up to Seattle looking for people who wanted to take a ride, and Bill Boeing paid the few dollars asked (in those days) to sample the experience. After the flight and a close study of the machine and its workings, he became convinced that a better device could be developed and built. He immediately set about achieving that objective. He hired some design and engineering staff, then presented the drawings they had made to his boat-builder who, together with some other specialists Boeing had employed, began the job of construction.

While this was in progress, to back up his own knowledge Boeing took flying lessons from a well-known aviator of the day, Glenn Martin.

The shores of Lake Union, just over a strip of land from Puget Sound and to the southwest of Seattle, were the baptizing waters for the tiny creation in June, 1916. The little float-plane was called the B&W (Boeing and Westervelt) Model 1. It was pushed out into the middle of the bayou for its first engine trials.

The test was an instant success, and Boeing decided to form his own aircraft manufacturing company, mainly to retain the services of the twenty-one specialists he had gathered together (besides his boat-builder). With the First World War raging away in Europe, he sensed that the importance of air power was not being fully exploited, but that when it was, there would be an urgent call for all persons knowledgeable in design and manufacture to put their experience to work. So, on July 14, 1916, The Boeing Airplane Company was born.

Prior to the end of the Second World War, the company concentrated on building planes for the military, though it also came up with a few notable civil airliners. History recalls many incredible World War II bombing stories involving the Boeing Flying Fortresses and Super Fortresses.

However, with peace, hundreds of orders for the armed forces were suddenly cancelled. The aircraft makers' production lines in most cases were choked with partially completed equipment that was no longer wanted. They had no new work. They had been stretched to the limit fulfilling the constant need for fighting machines. They either had to come up with radically different military designs—fast—and sell these to governments on the basis that the older models were then outdated, or they had to change their plans to concentrate on mass-transportation aircraft. But even there, most airlines found it far cheaper to purchase war surplus planes—sitting around in the tens of thousands—and alter the interior set-ups to make them a little more comfortable for their passengers. This meant that even new civil airliners would have to be bigger, faster, more reposeful, more economical, in order to interest the carriers.

Near the end of the war, the B-29 Super Fortress had been modified to the B-50, but peace had soon brought its production to a close. Boeing was searching for ideas in order to keep at least some of its work force gainfully employed. The company decided on three things:

1. To build a wholly new civilian plane with a huge double-deck fuselage, based on the B-50, and pre-designated the Model 377, the *Stratocruiser.*

2. To construct a jet-powered military bomber called the B-47, the *Stratojet.*

3. To work on updating their B-50 design so that it would take jet-turbine engines; this aircraft was later known as the B-52, the *Stratofortress.*

By April 15, 1952, all was in readiness for the first testing flight of their huge intercontinental bomber, the B-52. And just one week later, on April 22, 1952, the Boeing board decided to go ahead with another project it had been studying. This was for a jet-powered military tanker—an aircraft that could refuel the B52s in the air, thus giving them an enormous range. Because it was a speculative idea requiring an investment of twenty million dollars in "risk money," the board also decided it must be as versatile as possible in order to enhance its sales chances. Therefore, the design was such that it could fulfill the roles of (a) a military tanker, (b) a troop transporter, and (c) a civil airliner. The early drawings showed it as being coded the Model 707.

This machine, they determined, would weigh 190,000

pounds and it would be a low-wing monoplane. Its 128-foot length would contain a cabin that had, in effect, a flat floor (for demonstration purposes), but onto which carpets and between 80 and 130 passenger seats could be quickly arranged in banks. Its four pure-jet turbine engines and swept-back wings would allow it to fly at 600 miles an hour. At this stage, however, the Air Force was not interested.

On May 15, 1954, at 1600h, the huge Boeing hangar doors folded back; a band, especially hired for the occasion, started playing; and the large crowd pressed forward as a tiny (in comparison) tow-tractor pulled the immense bulk of the freshly painted new Model 707 out onto the surrounding apron and into the warm Renton sunshine. She was a bright canary yellow on top, with a copper brown, half-arrow-design "cheat line" running down the sides from cockpit to tail. Beneath this was the gleaming polished aluminum of her lower hull. "Boeing"—"Boeing"—"Boeing" was painted on her. And her wings and fin proudly displayed her registered number, N 70700. Even the company's special guest, 72-year-old William Edward Boeing himself, was amazed by its sleekness and its immensity.

Taxiing and ground static engine runs filled the design and testing crews' days for just a month. With (presumably) all the initial, superficial problems ironed out, the first flight was planned for 0800h on June 14, 1954. But when the morning came, the weather let them down. It was heavy, with a low overcast, threatening rain at any time. The hundreds of spectators who had amassed to watch—both those officially invited and others—were disappointed, but hopeful. At midday, the clouds started to lift. By 1400h, the sun was streaming. Test pilot "Tex" Johnson, wearing two heavy woollen jackets, a life-vest and a parachute pack, was strapped in his seat and the four Pratt & Whitney JT3C-6 powerplants roared away at full blast, eagerly awaiting the release of the brakes.

Nine months after this first flight, in March 1955, with

world trouble-spots coming and going and with the constant threat, so the Americans thought, of Russia's nuclear power raining down on them, the U.S.A.F. decided that its B-52 bombers should be capable of reaching anywhere on the globe non-stop. The Stratofortresses had been modified to carry H-bombs. The Air Force either needed a new machine altogether, or it needed a secondary one from which the B-52s could be refuelled. The Boeing Company's socalled "gamble" paid off. The Model 707 became the military tanker, the KC-135, even though it was then still in the testing stages.

At about the same time, the Douglas Aircraft Company was busily sounding out the airlines with a couple of drawings it had thrown together and headed "The DC-8 Project." As far as the civil side of the 707 design was concerned, Boeing realized it had the time advantage over Douglas. Its machine was already flying, even though the company knew that certain modifications might have to be effected in order to meet the legislated standards for jet-powered transports that were sure to come. And at about the same time, the British had uncovered the disastrous problems with the Comet (the world's first jet airliner, which had to be taken out of service and modified because insufficient testing permitted fatal faults to persist). They were then in the redesign stages of producing an even bigger and more powerful version, learning from the original mistakes not only the bad, but also the good: passengers liked the speed and comfort the new engines offered. The airlines that had scheduled the early Comets had been running them at full capacity most of the time.

The problem for Boeing was that the idea the Douglas people were offering, even though it was still only on paper and looked exactly like the 707, was bigger. Boeing would have to do something to increase its machine's size in order to compete. If it did not, it would rapidly lose out to Douglas.

The war of nerves between the two manufacturers ended in Boeing beating Douglas at its own game. Boeing, having no major prototype costs because the military had guaranteed most of them in its order for the KC-135 tankers, could undercut whatever price Douglas was setting. The DC-8 was quoted at $5,650,000, so the Boeing 707 came on the market at $5,350,000.

On October 13, 1955, Juan Terry Trippe, the president of Pan American World Airways, after being taken for a number of rides in the Model 707, announced that his company would be the first U.S.-based airline to introduce jet-powered aircraft. He signed an order for twenty-three of the planes. Pan Am also contracted for some Douglas DC-8s, knowing that they would be delivered much later.

Boeing was back in the civil airliner market.

With all the changes that were necessary to the KC-135, the 707 passenger prototype didn't take off into the wild blue yonder for its maiden flight until the afternoon of December 20, 1957.

Boeing still thought it was ahead of its rivals. The newly redesigned Comet, called "The 4," was not to be taken on its first air trial until April 27, 1958; the original DC-8 (which was in fact the initial production model) was still being built. In the meantime, the British government's civil servants, who had earlier certified the faulty Comet 1, had discovered a flight-controllability problem with the 707's rudder and said they would flatly refuse to pass the machine until it was modified. (Ironically, they were right, though their insistences were ill-defined, weak, and did not resolve the troubles.)

The Comet 4, flown by British Overseas Airways Corporation, entered passenger service across the North Atlantic on October 4, 1958. As a result of the necessary further design adjustments, the Boeing 707 Series 100, laden with Pan American customers, started on October 26, three weeks later.

During the first two years of airline operations the Douglas DC-8, also found to have many complications, caused its company to lose $109,000,000, with the development costs ending up in the region of $298,000,000. Boeing's development expenses came to about $165,000,000—thanks to the patterns and parts that were common to both the 707 and the KC-135 tanker.

As far as coincidences go, a similar accident occurred to both models during their testing stages.

The Boeing 707 was involved in a bad landing at the company's airfield near Seattle on April 17, 1958, 192 days before its epoch-making Atlantic crossing. It was coming in for a touch-down in a medium twenty-five miles per hour gusting crosswind when, as it neared the runway, it suddenly and rapidly started to lose height. The pilots pushed the throttles into high-power settings, but to no avail. The airplane descended hard on its port-main landing gear and the underside of the left outboard engine collided with the metalized pavement surface. An explosion occurred. The powerplant burst into flames. The machine was substantially damaged, but none of the four test personnel on board was injured.

The DC-8 started passenger service simultaneously with two U.S. airlines, United and Delta, on September 18, 1959. On May 14, just 129 days previously, a certification flight was drawing to its close at Edwards Air Force Base in California (which Douglas used for most of its aircraft trials). Both Federal Aviation Agency (as it was then called) and Douglas test people were aboard. An F.A.A. employee was flying the plane. He was using "minimum runway distance techniques" (stopping as quickly as possible after touching down). "Although the pilot precisely followed the [Douglas] recommended procedures," says the accident report, "the rate of descent near the ground was higher than expected by the manufacturer." The DC-8 crashed to the surface with such force that the fuselage was smashed apart and the outboard engine under the left wing was ripped off.

Luckily, none of the eight persons in it was injured.

But then the *real* casualties started.

The first happened to a nearly full Pan Am 707 flying at 35,000 feet above the North Atlantic. It was mid-afternoon on February 3, 1959. With the systems on automatic and the first officer busily working out their estimated arrival time in New York, the captain decided to take a stroll back into the passenger cabin to see how everything was faring. He had only been gone from the cockpit a few minutes when the autopilot suddenly disengaged itself and the 707 went into a steep descending spiral. It was just about to roll over on its back when the captain, assisted by some people in the first-class lounge, managed to struggle back to the flight deck. As he took his seat he noticed the altimeter pass through the 16,000-foot altitude mark. They were dropping fast, out of control, in dense clouds. The captain rolled the wings level and, as the plane sank below 8,000 feet, he pulled back as hard as he could on the yoke. Gradually the machine started to climb—6,000 . . . 7,000 . . . 8,000 . . . 9,000. Control had been regained and they continued normally, diverting to Gander, Newfoundland. In that extremely lucky experience, two people were seriously injured, and twelve others received minor lacerations.

Just twenty-two days later the next mishap occurred. A Boeing 707 was being used by Pan American for crew training purposes in the vicinity of Chartres, France. While flying on two of its four powerplants, they suddenly went into a right-hand spin. The instructor-pilot quickly took measures to stop it, bringing the plane back to a level attitude. But in doing so, the outboard engine under the starboard wing fell off, dropping heavily into a field below. The machine was safely landed at London's Heathrow Airport. None of the five persons on board was hurt.

The next event took place on March 26, 1959, when a repetition of the accidents that had been experienced during the final test flights happened to an American Airlines crew at Chicago. The 707, landing in a crosswind, came down heavily short of the runway. Only this time there were 114 persons involved.

And the same thing occurred *again* on April 16, 1959, when the trainee pilot flying a Pan Am 707 weightily touched it down before the concrete threshold at the Peconic Airport near New York.

On July 11, 1959, the design prototype, now registered as N 707 PA, took off from Idlewild, New York, as Pan Am's Flight 102 on its way to London. Inside were 113 people. At 2037h, just as it was lifting into the night skies, two of the four wheels on its left-main undercarriage detached. It was the tower controller who noticed them. They pounded back onto the runway's surface and, bouncing and rolling at a very high speed, shot off into the waters of Jamaica Bay. The Boeing remained in the vicinity for four airborne hours, using up fuel and permitting the airport to prepare for an emergency touchdown. It finally landed, successfully, at half-past midnight. But such was the fascination of jet-powered civil aircraft that a crowd of approximately six hundred massed around the stricken machine when it was safely brought to a stop. There was a risk of roller explosions, but the throng would not disperse, so fire hoses had to be turned on them.

During an investigation of the affair, Pan Am was blamed for improperly maintaining the airplane. This was not the first time that the accusation had been levelled against the airline—but it was the primary such charge involving its jet-powered fleet.

Another landing accident, with one hundred lives at risk, occurred on August 11, 1959. This time it was an American Airlines 707 at Chicago.

Then came the first major disaster.

The Boeing was practically brand new. It had been delivered to American Airlines on June 5, 1959, and had clocked up just 736 air hours when, on August 15, it ended up in a ball of flames three miles from the Peconic Airport near Calverton, Long Island. In it were a captain-instructor, two captain-trainees, a flight engineer-instructor, and a flight engineer-trainee. They had taken off from Idlewild (now J.F. Kennedy International Airport) at 1340h, and arrived in the Peconic area at 1511h. After flying in the vicinity for ninety minutes, they radioed the tower that they were making a landing approach. At 1641h, the plane suddenly crashed to the ground and exploded. Everyone on board died.

Much criticism has been leveled at the U.K. civil servants who certified the Comet 1 without adequately testing it. When this 707 disaster occurred, five years had passed since the Comet was taken out of service because of inherent design faults. This New York accident proved the American authorities to be equally as negligent as the British, possibly more so. The F.A.A. did absolutely nothing constructive to correct the inherent defects, even after they had been uncovered more than once.

Two months later, on October 19, 1959, the second major 707 disaster occurred, *again* on a training run. *Once more,* the aircraft went out of control. Of the eight persons in it, the four operating crew died, but the others—Braniff International Airways officials who were passengers—escaped with only serious injuries.

The plane had been on a "demonstration and acceptance" flight before being officially handed over to Braniff. Boeing thought it might as well gainfully use the air-time by also giving some practice to one of the airline's captains. Three of the manufacturer's test personnel were in the cockpit as well as the learner. The Braniff pilot decided to "bank" the airplane. The Boeing Company had earlier said that any operation of this nature should be restricted to fifteen degrees side tilt, with an absolute maximum of twenty-five degrees. The trainee leaned the plane over to an angle of between forty and sixty degrees. The Boeing check pilots just sat back and watched. All of a sudden the machine was out of control, with three of its four engines falling off and both wings afire. The crew thought they had the situation under control and radioed the airport that they were attempting an emergency landing in a field near Arlington, Washington. Seconds later, the airliner struck a group of trees and crashed.

As more and more 707s took to the air, thousands, tens-of-thousands, hundreds-of-thousands of passengers marveled at jet-powered speed and comfort. These latest Boeing models were the Series 320s, the Intercontinentals, complete with the more powerful, advanced, JT4A-3 *fan*-jet engines.

Over the following months, a whole chain of accidents occurred to the 707s in service, all during the landing stages of flight:

February 7, 1960 Pan American World Airways—substantially damaged landing at Los Angeles (cause: improperly executed approach by the pilot).

May 9, 1960 Trans World Airlines—substantially damaged landing at New York International Airport (cause: poorly conducted instrument approach).

June 19, 1960 American Airlines—substantially damaged landing at Dallas (cause: inadvertent actuation of the "stabilizer trim" at touchdown).

October 7, 1960 Air France—substantially damaged landing at Pointe-à-Pitre, Guadaloupe (cause: pilot error, touched down short of the runway).

October 25, 1960 American Airlines—substantially damaged landing at Fort Worth (cause: nose landing gear failed to extend, reasons unknown).

October 29, 1960 South African Airways—substantially damaged landing at Nairobi, Kenya (cause: pilot error in grounding 9,000 feet short of the strip).

December 24, 1960 British Overseas Airways Corporation—substantially damaged landing at Heathrow Airport, London (cause: pilot error, alighted too fast, too far down the runway).

January 9, 1961 Continental Airlines—substantially damaged by a final approach engine fire at Rockford, Illinois (cause: material failure in the port manifold).

Then, on January 28, 1961, yet *another* 707 went out of control during a training flight, crashing into the Atlantic off Montauk Point, New York. The investigators were unable to determine what caused the dive which preceded the impact; all they could find was about fifteen percent of the aircraft and its contents. It belonged to American Airlines, and the six crewmen it contained died.

Three days later, on January 31, 1961, a South African Airways Boeing—the same one, in fact, which had been involved in the landing accident at Nairobi a few months earlier—slightly overran the Zurich runway; its nosewheels collapsed after only they had left the far end of the pavement, next to which the Swiss airport's management had carefully designed and built a ditch. The Swiss inquiry into the matter had the impudence to blame the pilot.

Then it happened—the very thing that everyone knew eventually must. On February 15, 1961, a Sabena (the Belgian airline) Series 329 crashed near Bruxelles Airport, killing all seventy-two people inside it as well as one person on the ground, while seriously injuring another.

The flight had originated in New York and was on a long, gently descending approach to the airfield. As it neared the runway, instead of landing the pilot increased engine power and retracted the undercarriages. The aircraft started to gain height, then made several complete circles to port. As these continued, the 707 leaned further and further to its left until finally it was nearly in a wings-vertical position. It suddenly crashed to the ground and exploded on impact, erupting into sheets of flame.

The Belgian investigators were unable to determine the causes of this disaster. But, as in past cases, it was apparent that the pilots had lost control at a low speed.

A number of misfortunes of a minor nature then occurred to 707s. Most, as before, were caused by the pilots having difficulties during the final stages of a landing.

Until the next major calamity took place.

At Idlewild Airport at 1015h on the morning of March 1, 1962, an American Airlines Boeing was departing on a flight to Los Angeles with a crew of eight and eighty-seven passengers. The plane made what appeared to be a normal lift-off, and thirty seconds later began a gentle turn to the left. The pilots, leveling out, then continued with their climb for several seconds before starting another turn to port, having been instructed to do so by air traffic control. However, this time the degree at which the aircraft was banking progressively increased until, at right angles to the horizon, it suddenly flipped over on its back and plunged into Pumpkin Patch Channel in Jamaica Bay, practically disintegrating and bursting violently into flames. All ninety-five persons died.

As in the earlier, similar Belgian accident, no radio calls indicating distress had been received.

The search for a reason concentrated on the burned-out tail of the machine. A paragraph from the post-mortem report states:

"The investigation disclosed that the rudder servo wiring had an 'open' in the rate generator circuit. It was found that the brown wire, which connects the output of the rate generator to the input of the autopilot amplifier, and the orange wire which is the 'ground' or return side of the 18 volts input, were severed; and that the blue wire which connects 18 volts AC to the rate generator input, was holding together with only one strand. The separations of the wires were adjacent to one another. The nature and protected location of the wire damage precludes the possibility of such damage having occurred at impact. Also, *some spare servo units from American Airlines' stock and numerous servo units on the manufacturer's assembly lines* [italics added] were found with similar damage and markings. It was determined that the damage had occurred as a result of improper use of tweezers when tying the bundles of wire to the motor housing. This was considered to be conclusive evidence that the damage to the rudder servo unit on N 7506 AA was initiated by assembly and/or maintenance operations."

On June, 3, 1962, the death toll mounted again. Only two stewardesses survived out of 132 people on an Air France 707 scheduled from Paris to Atlanta.

The Boeing was taking off from Orly Airport and was up to speed when the pilot lifted the nose into the air. The plane sat there, pointing skyward, for between four and six seconds. Suddenly, the front undercarriage was brought back down onto the concrete. The brakes were heavily applied. Thick smoke streamed from the wheels. As they neared the end of the runway, the pilot tried to swing the aircraft around, looping it to keep it from running off. But it was no use. They raced at approximately 170 miles an hour over the grass beyond the end of the asphalt. Unexpectedly, in a patch of rough ground, the left-main landing gear gave way. The machine lurched violently to one side and the two port engines dug into the dirt. Fire broke out in the left wing.

The still fast-moving fuselage sped across the airport's perimeter road and as it did, the inboard left engine was wrenched off and the right landing gear collapsed. It then collided with the solid steel stanchions of the approach lighting system and started coming apart as it slithered down an embankment heading directly toward the river Seine. The broken-away forward section of the plane collided with a house and garage, while the remainder continued sliding for another three hundred feet before finally coming to a stop.

Nineteen days later, Air France was at it again. At 0400h on the morning of June 22, 1962, its 707 (which, ironically, had been involved in the descent accident at Guadeloupe on October 7, 1960) was letting down through clouds for another landing at Pointe-à-Pitre. The airport was under the control of the French authorities. They reported to the pilots by radio that one of the only two local navigational aid transmitters had suffered a malfunction. Possibly because of the insufficiency of instrument assistance, the aircraft, nine miles off course, crashed into hilly countryside killing all 112 persons on board.

Then on November 27, 1962, the Brazilian airline VARIG saw one of its 707s involved in an accident while it was banked to one side. It was coming around for a landing at Lima, Peru, and when the left-hand turn had been nearly completed, it suddenly smashed into the rock face of La Cruz Peak. It seems highly unlikely that the plane was out of control at the time; more probably the cause was pilot misjudgment.

Other destructive crashes involving Boeing 707-type aircraft ensued over the following years:

February 12, 1963   Northwest Airlines—near Miami—disintegrated in a thunderstorm—forty-three dead.
December 8, 1963   Pan American World Airways—near Elkton, Maryland—exploded in flight after lightning set the more dangerous fuel being used ablaze—eighty-one dead.
April 7, 1964   Pan American World Airways—landing at Kennedy Airport, the plane overran the runway into Jamaica Bay—forty-three injured.
July 15, 1964   Deutsch Lufthansa—near Petersdorf, Ansbach, Germany—out of control on a training flight—three dead.
November 23, 1964   Trans World Airlines—Roma Airport, Italy—the pilot thought he had troubles on takeoff; aborted, and collided with airport equipment; the more dangerous fuel being used exploded—fifty-three dead, twenty-three injured.
May 20, 1965   Pakistan International Airlines—near Cairo, Egypt—crashed short of the runway; Cairo notorious for lack of operable equipment—119 dead, 6 injured.
September 17, 1965   Pan American World Airways—near Monsterrat, West Indies—flew into a mountain in thick clouds—30 dead.
January 24, 1966   Air India—Mont Blanc, France—collided with the mountain after imprecise English was

used by a Geneva, Switzerland, air traffic controller—
117 dead.

March 5, 1966 British Overseas Airways Corporation—
near Mount Fujiama, Japan—disintegrated in flight—
124 dead.

At this period in its life, the 707 went through a phase of suffering an exceptionally large number of undercarriage problems. The landing legs either were not extending correctly in preparation for touchdown, or were collapsing on meeting the runway's concrete, and they were jamming, unable to be retracted from their extended positions after take-off. This pattern lasted for between two and two and a half years.

November 6, 1967 Trans World Airlines—near Cincinnati—the co-pilot thought a collision had occurred and tried to abort the take-off; the plane ran off the end of the runway—one dead, thirty-five injured.

January 9, 1968 Ethiopian Airlines—at Beirut, Lebanon—landed nose-gear first and extensive fire broke out—no injuries.

February 7, 1968 Canadian Pacific Airlines—at Vancouver—ran off the runway while landing and collided with an aircraft, three vehicles and two buildings—two dead, seventeen injured.

March 5, 1968 Air France—at Pointe-à-Pitre, Guadaloupe—collided with a mountain on a clear night—sixty-three dead.

April 8, 1968 British Overseas Airways Corporation—London's Heathrow Airport—engine disintegration soon after take-off; because crew used wrong drills, fire persisted; re-landed in flames—five dead, thirty-eight injured.

April 20, 1968 South African Airways—at Windhoek, Namibia—crashed to ground soon after lifting off; crew operated wrong handle—123 dead, 5 injured (though insurance figures show there were 7 more persons, all dead, on board; the South Africans have refused to supply a copy of their report).

June 3, 1968   Pan American World Airways—Calcutta, India—struck a tree a half mile short of the runway; crashed and caught fire—six dead, fifty-seven injured.

July 13, 1968   Sabena Belgian Airlines—near Lagos Airport, Nigeria—crashed into a tree eight miles before the runway; impacted and caught fire—seven dead.

December 12, 1968   Pan American World Airways—near Caracas, Venezuela—crashed into the ocean eight miles from the runway during a night approach—fifty-one dead.

December 26, 1968   Pan American World Airways—Anchorage, Alaska—destroyed, trying to take off without flaps extended—three dead.

July 26, 1969   Trans World Airlines—near Atlantic City, New Jersey—went out of control on a training flight after loss of tail rudder power during an hydraulics failure—five dead.

December 3, 1969   Air France—Caracas, Venezuela—soon after take-off, the plane's nose suddenly dropped; impacted with the ocean—sixty-two dead.

Up to this stage, the Boeing had been in passenger service for just over 132 months. In its first ten years, 757 aircraft had been delivered, including Series 120, 120B, 200, 320, 320B, and 420 of the Model 707 and also 720s and 720Bs, being a slightly shorter but otherwise similar design.

The 707 now on its way to Pago Pago—construction number 19376—was one of a batch of nineteen that had been ordered simultaneously by Pan Am. Of the others in that group, *Titan*—construction number 19368—had been mortally crippled in the freight crash at Logan Airport, Boston, on November 3, 1973, as we have seen. And *Rising Sun*—construction number 19371—had finished its days being flown into a hill about twenty miles from Manila Airport on July 25, 1971, while being brought down for a landing. It, too, was laden with cargo. The three crewmen died.

The Boeing 707 en route to Pago Pago was called *Radiant*. She had been delivered to Pan American on December 20, 1967, already tested and checked and, as not much could be found wrong, registered as N 454 PA. Whether she had the designed-in tail problems of all the earlier of her kind was not to be revealed, though she more than likely did. She had been decked out inside with 146 passenger seats—sixteen of them bigger, wider, more comfortable than the others—and soon she entered service, flying between most spots her master visited around the world.

Because she appeared to be well manufactured, her records showed few problems other than some minor ones. Naturally, she would spasmodically develop some aches and pains, but nothing terribly serious. During her work life she had had her regular checkups, and when she lifted off into the night skies over Auckland, it was the six-thousand-one-hundred-thirty-first time she had gracefully taken to the air. She had flown for a total of 21,625 hours, averaging eight hours and thirty-eight minutes of "flight time" each day. Though nothing had seemed intrinsically wrong, she had been laid up for a seventy-two-hour "heavy maintenance overhaul" at Miami ending on April 22, 1973. After all she had, representatively, performed two and a half landings every twenty-four hours of her existence. And her "statistical flying time" on each trip had been three and a half hours.

Her four powerplants, likewise, were in good condition. Her number one (outboard under the left wing) had been with her since February 22, 1972, and had recorded a total of 22,158 operating hours over 8,461 start-ups. The number two (inboard under the left wing) had pushed her along since April 11, 1973, with a total of 18,769 functioning hours and 6,181 starts. The engine in the number three position (inboard under the right wing) had been there since April 19, 1973, with a total of 22,744 hours and 7,373 starts. And the number four (outboard under the right wing) had been with her only since December 19, 1973, with recorded figures of 20,527 operating hours and 6,478 cycles.

Just prior to her take-off from Auckland, the engineers at the airport had made five minor repairs which had been written up in the flight log during the earlier Australia to New Zealand journey:

>*Log Sheet 852*—January 30, 1974.
>Radar light shield is missing.
>*Corrective action:* Found the shield and installed it on the indicator.
>
>*Log Sheet 853*—January 30, 1974.
>Replace flight recorder tape per equipment advisory.
>*Corrective action:* Replaced tape. Tests OK.
>
>*Log Sheet 854*—January 30, 1974.
>Please replace number 303 bulbs in bulb holder.
>*Corrective action:* Replenished spare bulbs.
>
>*Log Sheet 855*—January 30, 1974.
>Left-hand side fluorescent tube in "B" lavatory inoperable.
>*Corrective action:* Replaced tube.
>
>*Log Sheet 856*—January 30, 1974.
>Crew's oxygen system pressure on low limits.
>*Corrective action:* Topped up crew's oxygen. Valves locked "open."

So now, with her minor problems solved, her fresh crew, her cargo, her fuel, her passengers, she was once more in the air, running before the winds across the wide blue Pacific to tropical Polynesia.

5

The time was *2245h00*. Captain Leroy Petersen thought he had better remind the passengers that the cockpit crew was wide awake up front, so he announced over the public-address system that there were tropical showers at Pago Pago.

The news caused the Canns some consternation. Rain was the last thing they had been expecting; they had packed their waterproof gear into their luggage, which was now in the aircraft's cargo holds. Roger and Heather conferred and decided that once they had landed they would go straight to the hotel for the night, despite the excitement of their first international trip and the fact that sleep was not among their earlier priorities.

2304h00. Acting co-pilot James Phillips again talked by radio to Nadi, reporting that they were 150 miles from Samoa and still at their 33,000 feet cruising altitude. He was advised to continue the journey toward the airport.

The Fijian operator teletyped his counterpart in Pago Pago, checking that all was well for the plane's imminent arrival.

2311h00. Nadi flight control told Phillips to change his radios over to 121.0 mHz so that he could speak directly with Samoa.

2311h55. Once the tuners had been reset, the co-pilot switched on his microphone: "Pago Radio. Clipper Eight Zero Six."

"Clipper Eight Zero Six. Pago Approach," came the reply.

"Roger, Pago Approach. We are now one three zero miles south. Can we have the latest weather there, please?"

"Clipper Eight Zero Six. Roger. The Pago weather is: ceiling, estimated at 1,600 feet, broken; 4,000 feet, broken; visibility . . . ah . . . correction . . . 11,000 feet, overcast; the visibility is 10 miles; light rain showers; temperature 78 degrees; wind, three five zero degrees at 17 [miles per hour]; and the altimeter is two nine eight five [inches of mercury]."

"Roger," replied Jim Phillips. "Ah . . . that's wind at three five zero, at one seven, huh?"

"Variable to the north-northwest here. About three four zero degrees at one seven."

"OK. Fine. Thank you."

"And Clipper Eight Zero Six. You're cleared to the Pago VORTAC from your present position, via 'Ladyfish' 'Foxtrot' route. Ah . . . descend and maintain 5,000 [feet]. Report leaving altitude three three zero. And report by 'Ladyfish'."

"Understand. We are cleared to the Pago VORTAC via 'Ladyfish' 'Foxtrot'. We are cleared to 5,000."

"That is correct," replied the controller. "Report leaving three three zero. And report by 'Ladyfish'."

"Roger," Phillips acknowledged.

The flight crew retrieved their charts from the different hiding spots next to seats or on top of ledges on the equipment. They searched them in the marked vicinity of the airport's immediate area for the placement and frequency of the VORTAC radio navigation beacon that would give them, on their instruments, a compass reading from the transmitter sited next to the far end of the runway.

Another aerial map was consulted so that they could see the position of the air route crossroads called "Ladyfish." They further needed to know the direction to be taken to fly along the imaginary path leaving the intersection and named "Foxtrot."

Pago Pago International Airport is a part of the forgotten backwoods in the American Federal Aviation Administration's system—far away in both distance and mind from the luxury of the F.A.A.'s large, marbled, air-conditioned Washington, D.C. headquarters.

It is on the central southernmost coast of the island of Tutuila, with just one concrete-based, asphalt-topped functional strip, 9,000 feet in length and 150 feet wide. (These figures conform with the international standards laid down as acceptable for runways intended for long-haul aircraft.) Planes using it must approach from the south-southwest for a landing, and take off from the north-northeast. This is necessary for two reasons: first, because of the relatively close range of the mountaintops on a nearby island; and second, because the airport has only one set of radio landing aids.

However, it is furnished with high-intensity runway lighting, a medium-powered lead-up illumination system, runway alignment lights, and a visual approach slope indicator system commonly called a VASI. Impressive as this may sound, it does not mean the airfield could be classed in any way as "well equipped" when compared with most others. There is, on top of this, one inherent problem in that tropical region: maintaining continuity and regularity of the electricity supplies. Pago Airport has suffered innumerable power problems resulting in the failure of ground-to-air transmissions during crucial flight stages—more times than not amid the flooding equatorial downpours so typical of the area.

Besides this, the torrid climate and the local terrain played havoc with the electronic landing signals (not to mention the fact that they were misaligned to begin with), "bending" them out of their straight skyward course. This missetting of the angle was well known to the F.A.A. and its technical personnel. The flexion problem was nothing new either, and it was extremely difficult to detect when it was taking place. The changing position of the radio beams, combined with other factors, had been proven to be a major contributory cause of a number of accidents, notably to a Japan Air Lines DC-8 that gently "landed" in San Francisco Bay on November 22, 1968. In addition there were four major disasters (and nearly a fifth to a Qantas Boeing 707) all at New Delhi, India, when, within a year and a half a Japan Air Lines DC-8, an Indian Airlines Corporation Fokker F-27 Friendship, an Indian Airlines Corporation Boeing 737 and a Lufthansa Boeing 707 were all wiped out, killing a total of 152 persons.

Pago Pago International Airport had been certified by the F.A.A. on May 21, 1973. The documentation noted that the landing field failed to meet Federal Aviation Administration standards in only one respect: it did not have a perimeter fence to keep sightseers and straying animals away from the equipment, runway, and aircraft. The Government of American Samoa promised faithfully that the enclosure would be erected no later than November 20, 1973. So far, it had not been done.

Because it was a small, infrequently used place, the single air traffic controller on duty sat in an office in the main terminal; his only view outside was through a window at the far end of the room. If he leaned back in his chair he could see just a part of the runway's edge where it disappeared over a slight mound. He could see a bit more if he moved his seat back a pace or two, or if he stood up.

His desk was connected by teletype to the international aviation communications system, and to the Pago National Weather Service located in another building on the airport's grounds. He had one radio transmitter/receiver on a fixed wavelength that not only put him in contact with incoming and outgoing flights, but was shared by the Port Management's staff. It was also his main life line to the fire and dispatch headquarters. The brigade's duty officer, on another frequency, was in radio contact with the rescue trucks when they were out in the field.

In front of the controller as he awaited the infrequent arrivals and departures was a series of "dimmers" (which had been made inoperative some time ago) for varying the intensity of the approach, runway, and taxiway lights; wind speed and direction indicators; and a specialized barometer for taking precise air pressure readings.

Because of the lack of control tower facilities, one of the airport's five fire trucks was stationed halfway along the strip when an aircraft was coming in or taking off. As the traffic director had no clear view of the runway, should an accident occur, the driver of the waiting vehicle, theoretically, would immediately radio the alarm to his base for further assistance.

The divisional reports told their own stories. For example, during December 1973, though there had been 203 commercial and 34 military flight movements, fire and rescue headquarters had been called just once. The single emergency was to attend a blaze in the heavy equipment compound; it had been brought under control by the time the bowser arrived. In that same month, there was one rescue operation. A Korean junk had become stranded in the water near the end of the old disused runway. The owner, a Lee Chung, had suffered minor injuries, but his craft was reported to have been in good shape.

Things had picked up for the service in January 1974. There had been a false alarm at Leone. The rubbish around one of the airport's incinerators had ignited and had been quickly extinguished. And a fork lift truck had burst into flames which, too, had been rapidly put out.

So at 2315h that evening, the standby engine, numbered

R-1, made its way to its usual position next to the runway for the arrival of the two-hundred-ninth commercial aircraft movement of the month, Pan American's Clipper 806.

The truck was driven by Fiaoo Masina, a forty-seven-year-old senior fireman. He was accompanied by his assistant, twenty-year-old Laalaai Niko. Both were Samoan of many generations. Neither spoke English.

On the way out to their stopping place near the windsock, Masina commented to Niko that, from the looks of the clouds above, the wind would soon start to pick up and they could expect a heavy tropical downpour.

When they reached their position Masina turned off the vehicle's motor, and they sat relaxed, waiting.

2316h58. Back in the Pan Am Boeing 707, James Phillips again radioed the Pago controller: "Clipper Eight Zero Six is out of three three zero for 5,000 [feet]."

"Roger."

Phillips must not have heard the brief acknowledgment. He called again: "Pago. Did you copy Clipper Eight Zero Six?"

"Clipper Eight Zero Six. Roger."

A further seven trouble-free minutes ensued.

2324h40. "Clipper Eight Zero Six. We are by 'Ladyfish,' out of flight level two zero zero for 5,000 [feet]."

"Clipper Eight Zero Six. Roger. And Clipper Eight Zero Six, you're cleared for the ILS-DME Runway Five approach via the two zero miles arc south-southwest. Report when making the arc, and when leaving 5,000."

"Cleared for the ILS-DME via the twenty-mile arc. And report out of 5,000. Clipper Eight Zero Six."

"Roger."

Again the charts were consulted so that the crew could jot on their note pads the radio frequencies of the combined Instrument Landing System's signals and the Distance Measuring Equipment's transmissions. They would soon need to dial these settings on their different telecommunicative and navigational instruments so as to be led down to the airport's runway.

Another five minutes of quiet work-load successfully passed.

2330h32. James Phillips uttered into his headset microphone, "Eight Oh Six is out of eleven for five."

The sudden spurt of life in the broadcast receiver caught the controller off guard: "Eight Oh Six. Pago. Say again!"

"Out of 11,000 for 5,000 [feet altitude]."

"Roger."

"And ask him what the wind is now," suggested the pilot-in-command.

"Ah . . . Pago . . . What is your wind?" the co-pilot inquired.

"It is three six zero, variable to zero two zero degrees now; one one miles per hour, gusting to one seven."

"Thank you."

"It's coming around even more," Pete Petersen remarked to his crew.

"Yeah. But you watch. It'll clear off and be absolutely calm."

"Sure."

"Just before you land, you'll get a twelve m.p.h. gust on the nose. That should let you down pretty lightly."

The four men in the cockpit chuckled. However, it was a rather tense, nervous laugh that came from Captain Petersen. Even though he was a highly experienced pilot with more than 17,000 hours of flying time to his credit, the only landing he had ever executed at Pago Pago International Airport had been during daylight on May 16, 1973, more than eight months before. Jim Phillips, on the other hand, had handled seven take-offs and nine landings there since October 11, 1973, within the previous fifteen weeks. Petersen, nevertheless, had elected to make this touch-down himself, with Phillips monitoring the approach and the telemetrics.

2331h21. "Well," declared the boss, "I'll follow my needles [his instruments] and I'll put 68 in the window [put his landing-speed reminder 'bug' at 168 miles per hour, their calculated touch-down rate, including an allowance for the reported head wind]."

Someone mumbled something unintelligible and Phillips replied, "Well, the temperature's seventy-seven degrees now. We should be in pretty good shape."

2331h56. The flight continued on its pre-set heading. Every now and then it was being gently blown off course in one direction or the other by the light atmospheric puffs outside. During this part of the journey, slight corrections were being made to compensate for them. At times the deviations were combined with a medium shuddering as the machine passed through a stratum of overcast on its way down.

The captain turned to Phillips and asked, "Have you got a flashlight handy there, Jim?"

The co-pilot nodded.

"Can you hand it to me? I just want to see what those windscreen wipers look like."

Phillips passed it over.

As Petersen stretched forward in his seat he must have touched it, in the slight jerkiness they were experiencing while descending through the cloud shelves, against the window. There were some clearly audible metal-to-glass clicks.

2332h09. "The heat's off?" The captain was inquiring whether all the aircraft's systems had been shut down. They were entering the hot, dank, tropical lower troposphere. Gerry Green, the flight engineer, switched off the warmers.

"Twenty-six miles out," confirmed Petersen.

Someone grunted.

"I'm gonna sneak over there a little bit," he added, probably pointing.

2333h35. The tense silence of the last minute was interrupted by co-pilot Phillips: "OK. You're twenty-one miles out."

After a two-second pause, Petersen stated, "OK. Fine. We're now twenty miles out from the DME, starting our arc."

"Roger," Phillips replied as a sort of double-check on the pilot's calculations, adding, "He just wanted a call at 5,000?"—reminding himself and confirming with the others that their next radio contact with the Pago controller was to be at that altitude.

It was at this point that the pilots turned on the "Fasten Seat Belts" signs in the passenger compartment—an indication to the crew and the others on board that they were on their way in to the strip. The signs also acted as an initial "prepare-for-landing" warning. Promptly the main cabin lights brightened and the senior purser, Gorda Rupp, announced over the loudspeakers that they would be touching down in about fifteen minutes and asked everyone to start preparing by complying with the now-lit instructors.

Heather Cann immediately placed her handbag beneath the seat in front of her, and she and her husband both buckled their belts and drew them as tightly as they would go around their waists. Along with many of the other travelers aboard, Heather then gazed through one of the windows in an attempt to discover what the countryside looked like beneath them. But it was a clear, dark, moonless night outside and, although they didn't realize it, they were still over the wide, wild Pacific. They would remain so until they had completed the twenty-mile arc and were plying toward the headland jutting out from southern Tutuila Island.

Looking back inside, Heather watched the stewardesses as they walked along the aisle, verifying that everything, including seatbelts, was in place. As is usual on these long-haul journeys, the main lights in the passenger compartments were dimmed once more, making it easier for the people to see the landscape below now that the airport approach was beginning.

Up front, the tense cockpit silence continued as each crew member checked and rechecked the readings on the dials. Minor directional adjustments were being made as the wind's light gusts continued to pat them slightly off course. Then, as an indication that the pressures were easing, someone started humming "Show Me The Way To Go Home."

2334h16. Phillips realized at the point when 6,000 feet appeared on *his* altimeter, that they had not yet made a comparison for agreement between his and the pilot's instruments. He asked, "A thousand to go on yours?"

"Right," came the reply.

"Assigned altitude of 5,000."

"And the VOR is starting to come in. I'm going to navigate automatically on that."

The twenty-mile arc progressed. As the aircraft continued to bank around, the Canns, in the passenger cabin, caught a glimpse between the now-thinning clouds of some pinpoints of light on the island below. Moreover, Heather realized they were turning because the intermittent twinkling specks suddenly became hidden from view behind the wing and she had to await patiently their reappearance a few seconds later.

Roger Cann instinctively knew that the nose of the big 707 was inclined, that they were descending. And then, for no obvious reason, he felt them leaning to the right and thought they were making a turn in order to try for *another* landing attempt, the original not having been completed.

2334h34. A "buzzing" sound reverberated around the cockpit. It was the altitude warning device, reminding the pilots that they were fast approaching the height which had been set previously on its digital dial.

"Call him and tell him we're 5,500, starting the DME," suggested Captain Petersen with a deafening cough that practically split his sentence in half.

Phillips again flicked his microphone key: "Clipper Eight Zero Six is out of five point five for five. And . . . ah . . . intercepting zero four eight . . . ah . . . zero four six degrees radial from the DME."

"It's the twenty-one radial really," the captain prompted him.

"There's the two two six radial," said Phillips, pointing to the figures on the radio compass.

"Eight Oh Six. Right," came the voice of the airport's controller. "Understand you're inbound on the localizer. Report about three [miles] out. No other reported traffic [near the landing strip]. Winds, zero one zero [degrees] at one seven gusting to two four [miles per hour]."

"Eight Zero Six."

2335h26. "To the ILS now?" asked Phillips, inquiring whether he should retune the receivers.

"All righteeee."

And as the radios were reset, the continuous pips of the instrument landing system's identification signal could be heard in all their headsets: "Dit-dit. Dah. Dit-dit-dah. Dah ['ITUT' in Morse code]."

"Report how far out?" Petersen was wanting to know when the next verbal contact with the airport would be required.

"Ah . . . three miles," Phillips replied.

"Three miles?"

"Here it is." Someone presumably held up his note pad for the captain to see. "Report three."

"OK. We can go down to 2,500 now then, Jim."

"OK. And you're cleared for the approach."

"Yeah."

"OK. The ILS identifier is on number one [radio receiver], huh? And zero four zero [degrees] should be set in the window," Phillips said cautiously, making sure that they had aligned their instruments correctly.

"Yeah."

"And the altimeter . . . " The co-pilot's voice trailed off as the microphone swayed from his lips while he leaned forward.

"Please give me the distance every once in a while," the captain requested.

"OK. One three, going on for one two."

"OK."

At this point they were still slightly more than twelve miles out from the Pago Pago airport. The ride inside the 707 began to become just a bit bumpy, as it normally does when it approaches and enters the tops of the lower-altitude clouds. The gentle shaking started at 2335h15, with the aircraft at a height of 5,150 feet. They reached the 5,000-foot level at 2335h32, coinciding with the end of their twenty-mile arc, and two seconds later the crew cut back slightly on the engines' power in order to reduce the plane's speed from 298 miles an hour.

2336h11. "Flaps to fourteen degrees," instructed the captain.

Phillips moved the "flaps operation handle" on the central console near his left knee into the fourteen-degrees-extension position. The huge slabs of metal, hidden most of the time inside the rear internal sections of the wings, slowly, deliberately started to wind themselves out, angling downwards.

"And the pressure altimeters?" asked the flight engineer.

"Set at two nine eight five . . . right and left," the co-pilot confirmed, checking as he did that both his and the captain's instruments were correctly on the same barometric reading, the one radioed a few minutes earlier by the Pago controller.

"Five thousand feet." Someone was reminding the pilots that they were at their holding altitude. Their heading was correct on zero four seven degrees, but by then their velocity through the air had been reduced to 248 miles an hour.

2336h25. "Landing bugs?" The engineer continued his approach-check-routine.

"Set at . . . ah . . . fifty-five and seventy-two . . . right and left," responded the co-pilot, indicating that they had taken into account the added speed needed to compensate for the gusting head wind which they had been warned about but which they had not as yet met.

"Distance?" asked the captain.
"Distance coming up on eleven."
"Give me flaps at twenty-five degrees."
"Two five," replied the co-pilot as he pushed the flaps operation handle one further notch along its gear-levered slot. The huge pieces of metal extended even further out and downward from the backs of the wings.
"And gear down."
Phillips moved the "undercarriage operation handle" from the "retract" to the "extend" position, watching his panel as the three green control lights flickered to red, confirming that the landing legs were no longer stowed and locked, and therefore that their powerful motors were slowly winding them down.
Heather Cann was still blindly searching for something of interest out of the window. The few-and-far-between specks of light on the sparsely populated, dark world below were there one moment and gone the next. She turned her head toward her husband, then suddenly noticed the hissing and whirring sounds of the undercarriage bay doors opening, permitting the rushing air to swirl into the now-exposed landing gear housing. She then heard the gentle but aggressive purr of the motors as the wheels were driven down from inside the aircraft's belly and stretched earthwards into the onrushing atmosphere. The two separate noises made her think the legs and been lowered and instantly raised again. Even though this confused her, she said nothing about it.
2337h04. "OK. We've got three greens. And we have the pressure," commented Phillips, announcing that the return of the emerald dots meant the landing gear had fully extended. But at the same time, with some puzzlement, he also noted a red light indicating a possible loss of hydraulic compression in the system. At this stage it was not too serious a matter because the three greens showed that the wheels were locked in their down positions.
"OK. Apparently it's the microswitch," the engineer re-

ported after checking the hydraulic meters on his control panel. Adding, "I hope."

Everything suggested that there was no pressure loss. More than likely it was simply another malfunction in a warning circuitry.

"OK," said the captain.

The engineer continued his checks: "Seat-belt signs are on."

"No Smoking," Petersen ordered.

Gerry Green turned to his control board and switched them on too.

"Approach check list is complete. Landing gear . . ." and the engineer's voice trailed away as he leaned over to his other panel, making sure that the red hydraulic alarm light had gone out.

"Flaps to forty," Captain Petersen commanded.

"Down. Three greens. We have the pressure." The copilot's eyes were on the undercarriage indicators and the "pressure lost" warner which he confirmed was no longer lit.

"Three greens. Four anti-skid releases. 'No Smoking' is on." The engineer was again visually scanning the undercarriage, braking system, and the passenger cabin signs.

"Flaps to four zero," the co-pilot added, indicating that the pointer showed the panels on the backs of the wings had now extended to their next sequential position.

"Stand by on the wipers for me, Gerry," instructed the captain. Possibly he had noticed the first tiny specks of rain on the windshield, or possibly it could have been a preparatory caution, remembering the reports of showers up ahead.

"Right," came the reply.

2337h42. "Arm instrument warnings." Gerry Green had started his next pre-landing drills.

"Armed," indicated the captain.

The engineer continued to read down the long, unavoidable, unforgivable list in his Operations Manual: "Speed brakes?"

"Forward," Petersen replied. "OK. Distance check."

"OK. You've got eight." Phillips had obtained the figure from the distance measuring equipment; now, looking up, he added: "I have the runway in sight."

2338h03. "We've come up on 2,000 on the radio altimeter," reported the co-pilot.

(At this exact point the flight recorders show they were actually at 2,150 feet; their heading, which moments earlier had been slightly to the left of the extended landing strip's centerline, had moved two degrees to the right; there was still the normal amount of gentle bumping around as the aircraft came closer to the ground, forging its way toward the field across the southern, steeply hilled coasts of the island, and their speed, which fifteen seconds before had picked up from 211 miles an hour to 217, had dropped back to 203, reducing toward the 168 m.p.h. approach velocity they needed.)

"Seven to the DME?" asked the pilot as he moved the throttle handles slightly forward, increasing the engines' revolutions in order to retard both their pace and descent losses.

"Seven. Coming up on seven miles."

"Check," said the captain, indicating he wanted the co-pilot to tell him when they had arrived at the seven-miles-to-go point.

"Two thousand," called out Phillips. (The recorder shows at that moment they were slightly below that height.) "Seven miles. And the altimeters check," he added, indicating that the height finders gave identical readings.

2338h28. The pilot again slightly increased engine power, this time to hold the aircraft at an altitude of 1,750 feet.

Directly in front of them, and just less than 9,000 feet from the start of the landing strip, was a tall outcrop of rock, Logotala Hill, with its flashing red hazard beacon standing 390 feet above the runway's surface. To let down prematurely would mean ending up in the wrong side of the hill. Not to let down soon enough would mean over-flying the airport altogether. For the next two minutes, the aircraft's altitude would be critical.

"A little bouncy out here," commented Petersen as they recrossed the coastline, bounded by 200- to 300-foot-high slopes.

A series of eight clicks suddenly was transmitted over the radio. They sounded like someone activating and releasing his microphone button. They originated, either accidentally or on purpose, from one of the ground station positions, the people patiently awaiting the arrival of the Boeing 707.

2338h51. The device sparked into life once more.

"Clipper Eight Oh Six. Pago. It appears that we've had a power failure at the airport."

"Eight Oh Six. We're still getting your . . . ah . . . VOR . The ILS and the lights are showing."

"Can you see the runway lights?"

"That's Charlie," meaning they certainly could.

"Ah . . . we have a bad . . . ah . . . rain shower here. I can't see them from my position here," the controller agonized.

"Ah . . . OK. We're five miles out from the DME now and . . . ah . . . they're . . . They still look bright."

"OK. No other reported traffic. The wind is zero three zero [degrees] at two two [miles per hour] gusting to two eight. Advise me when you are clear of the runway."

"Eight Zero Six. Wilco."

The rain was teeming down in the form of a localized tropical storm, but it was confined at this time to the far end of the landing field where most of the terminal buildings were situated. The 707's speed, height, and heading remained constant over this section of the flight. (Speed, 184 miles an hour; height, 1,770 feet; heading, a gentle rocking from side to side between zero four seven and zero four nine degrees.)

The heavy downpour was passing slowly across the airport from the northeast to the southwest. During the final stages of the Clipper's approach, it was confined to an area stretching from the far end of the field to a point little more than halfway along it.

Fiaoo Masina and Laalaai Niko, in the fire and rescue truck parked alongside the asphalt-topped strip, commented to one another that the rain was making it almost impossible to see the runway lights only yards away.

In the airport's coffee shop, James Prendergast and his two companions, along with everyone else nearby, had to keep shuffling their chairs and table back away from the large open window spaces to keep themselves from getting wet as the strength of the storm increased.

Farther to the southwest where the Pan Am 707 was coming in, the situation did not even warrant the windshield wipers being turned on as yet.

None of the passengers noticed any signs of rain on the windows beside them. Dick Smith, in seat 20-F, was looking out at the time. But because he was on the right-hand side of the plane, his view was still directly over the dark Pacific and therefore he could see nothing of interest.

Roger and Heather Cann, on the left side, noticed a widely separated row of glimmering pinpoints which might be a road. Roger, in the center of the three seats, was peering over his wife's shoulder trying to look at the available view. He checked his watch. Either the flight was running early, he thought, or Samoa must be a fairly expansive island. According to his calculations, they were not due to land for another eleven minutes. When he returned his attention outside once more, he too could clearly see the shimmering dots below. They appeared to him to be two or three street lamps—they were stationary—and then several pairs of what seemed to be moving lights. Automobiles, he guessed. He also noticed one fairly large illumination coming from a house through what may have been a French door—it was certainly bigger than an ordinary window. He could see it all very clearly. And if there had been people inside that well-lit room he could have seen them too, he thought. He knew that this was probably an impossibility because of the height at which he presumed they were flying, but the clarity, the transparency of the midnight atmosphere was striking.

About this time both Dick Smith and Roger Cann felt the airplane "fish-tailing," gently swinging from side to side. It was nothing severe, just the normal motion caused by pilot adjustments whenever turbulence blew them slightly off course. Neither was concerned about it. Roger Cann was faintly intrigued, since this was his first time in the air. The movements had drawn his attention to the inside of the cabin. Yet all he could see were the seat tops and the backs of the other passengers' heads. The stewardesses evidently were already positioned and awaiting touch-down, their final checks having been made.

Because there was nothing of interest to watch out his side, Dick Smith leaned forward, lifted up his briefcase and rested it on his lap. After searching about inside it, he removed his passport, his flight-ticket stub, a yellow slip he had filled out because he was staying overnight in Samoa, and an envelope containing a couple of letters of introduction. He also took out about $2,000 in cash he had been carrying with which to buy some videotape equipment when in Japan; it would help him, he thought, with his diving instruction. He put the landing card in his passport and placed them all, including his ticket, in the breast pocket of his sports jacket. Then he sat still, thinking for a moment. No, it would be better if the money remained in his briefcase, he decided. He removed it from the pocket, put it back into his portfolio, closing it and returning it to the position beneath his seat.

Meanwhile, the airport was slowly looming larger—closer. And so was the crest of Logotala Hill, 1.7 miles before the runway.

2339h44. "Keep your eyes on it. I'll stay on my instruments right here." The captain was telling the co-pilot to continue watching through the windshield, making sure the fast-approaching landing field and its lights remained constantly in sight.

"All right, sir."

"Keep your eyes on it."

They both knew that if they lost visual contact it might be because of a sudden drop in altitude; the glow from the airport being hidden behind the hill.

"I'm still on the VOR."

"Yeah," asserted the captain. "Wipers, please."

The aircraft, now nearing the advancing edge of the storm that was drenching the far end of the strip, had encountered, at worst, a very light misty rain. But it was nowhere near heavy enough to block out their view of the runway just two miles ahead.

The wiper motors started up at exactly 2339h56, the grunt-thud grunt-thud of their movement easily discernible in the nerve-charged cockpit atmosphere.

2340h02. "You only have your flaps at forty," commented the engineer. He was reminding the pilots they were not all the way down.

"That's right."

"Flaps where?" Phillips asked.

"Flaps to fifty."

Phillips leaned over and pushed the handle on the central console into the "fully extended" position.

"OK. You're over Logotala." Gerry Green was relieved to announce to the crew that they were straddling the hill, and now all that was left was to get the plane down onto the concrete.

2340h17. "Let me know when you've got the runway," Petersen said to Phillips.

"You HAVE the runway," he instantly, emphatically replied, "but you're a little high."

The captain eased the flight control stick forward.

"Flaps going to fifty."

The radio altimeter warning system gave an acknowledging "buzz," reminding them that they were losing height—fast—and were now close to the ground, though the change was not severe enough to elicit any human comment, critical or otherwise.

"Get your power up," cautioned someone with an eye on the instrumentation.

"OK. One seventy-two." Phillips' attention, moving from one read-out to another, was referring to the fact that they were nearly at their previously computed landing speed.

The pilot again increased the engines' revolutions.

Moments later, the altimeter warning "buzz" stopped. Almost immediately, it started again.

"Four hundred feet."

2340h34. "You're at minimums," the co-pilot stated with a note of urgency in his voice. He had been watching their height and the radio compass which indicated that their heading had changed from the required zero four seven to zero four nine degrees. Then attention had moved to the speedometer. It was reading 169 miles an hour—*and falling.* And all the while Clipper 806 was being slightly shaken by the turbulent winds.

The captain once more added engine power.

2340h36. "Field in sight." Phillips was *again* reassuring them that the runway was directly ahead, looming out of the tropical darkness, now nearly one mile away.

2340h39. "Turn to your right." He realized the airplane's nose was pointing about four degrees to the left-hand side of the center of the approaching strip.

2340h41. A sudden, unexpected, loud rustling noise filled the forward section of the passenger cabin. Dick Smith's attention was immediately drawn to it. It seemed to be something scraping or brushing along the underside of the fuselage. He also thought he felt the machine suddenly lose height. He sensed all was not well. In shock, his subconscious started talking to him: *"Hang on, buddy,"* it whispered, *"we're going to crash."* He quickly prepared himself for possible impact by tightening his belt, leaning forward, and grasping the sides of the seat in front of him. Just before he bent down, he glanced out through the window. He could see absolutely nothing.

The scrubbing, grazing, scratching sounds increased in intensity. Smith, his head inclined, had never heard anything like *that* before, never, not on a single one of the hundreds of other flights he had taken.

Roger Cann also noticed them. They appeared to him to be coming from somewhere near the front of the airplane. They were certainly weird. But he accepted the clatter as just another strange feature of what he decided was a normal landing process. He remained relaxed in his position. Everything seemed to be going well.

Heather Cann was again watching out of the window, so intently now that she did not hear anything unusual. She was thinking, as she took in the limited view, what a nuisance the rain would be, interrupting their plans for a midnight stroll along what she presumed would be the dimly lit, palm-lined streets of a small tropical town. Maybe the forecast was wrong; no rain could be seen as yet. And all the while she was fascinated by this road with the widely separated cars running along it. Then, every so often, her attention was drawn to flashing on one of the shiny metal engine pods dangling beneath the wing—the reflection of the red anti-collision light rotating on the underside of the fuselage. For no specific reason she glanced back toward her husband. Nothing seemed unusual. The other passengers were all calmly seated, evidently awaiting the familiarity of the wheels touching down on the concrete.

2340h42. "One hundred and sixty-two!" The co-pilot's voice rang with urgency. He wanted to draw the captain's attention—IMMEDIATELY—to the fact that they were *under* their computed minimum landing speed. Both pilot and co-pilot apparently were concentrating fully on their instruments, the latter regularly throwing a glance out of the windshield to make sure that the lighted strip was still visually ahead of them.

2340h43. Just two seconds after the scratching noises had begun, Clipper 806's nose landing gear gently sank into the soft earth 3,628 feet short of the runway. In those two seconds the aircraft had traveled 236 feet from the point where it had clipped the top of a Eucalyptus to where its left wing smashed through a palm tree and, 49 feet further on, it touched down. It then almost immediately "bellied" into the dense tropical jungle, cutting a straight swath 148 feet wide through the saplings, brambles, and undergrowth. At the same time, it started shedding some of the lighter pieces of metal from its air-conditioning bay doors and the flaps at the backs of the airfoils.

Just as the plane made its second caress of the jungle, Phillips looked up. He was shocked into silence. Seeing dimly what was happening in front of them, he grabbed the flight control stick and pulled it back hard, hoping for the miracle that would return them to the air. It was no use. So he instinctively leaned over and pushed the four throttle handles on the central console into the maximum positions, praying the powerplants would quickly wind up, spewing take-off thrust behind them and, with any luck, pushing them up and away.

The passengers felt the nose wheels make contact with the earth. A fraction of a second later, the two engines on the right side started digging eighteen-inch semicircular furrows into the soft tropical soil.

Heather Cann, who was still looking over at her husband, assumed that the slight jolts were simply part of a routine landing. Roger thought so too, especially since there had been no sudden change in the aircraft's direction or speed. In a crash situation, inexperienced in flight as he was, he would have expected incontrovertible violence. So he assumed that the plane had landed on the runway, and his jittery nerves were relieved at the relative gentleness. Then, when the shock of the nose gear hitting the ground was transmitted to his brain, in that split second he realized all was not well, that they were probably in deep trouble.

Yet, at the same time, he was aware of the calmness of the other people around him. So he simply waited attentively, in a mental quandary, to see what would happen next.

The sense of impending disaster came intuitively to his wife Heather. She immediately swung her head around and glared out of the window. She could see no rain on the outer pane. She could see no road, no cars, no house. Instead there were flames, huge balls of fire, gushing up from somewhere back of the wing, close to the fuselage. Yet even though the blaze was a short distance away, separated from her only by the light metalwork of the airplane's skin, she never for an instant thought that they might not escape from it.

The moment the nose wheels struck the ground, Dick Smith raised his head from his braced position and stared out of his window. He could see flames on his side of the aircraft, too. They were spurting up from somewhere just in front, over the leading section of the right wing. It looked to him as though the inboard engine was on fire. While he watched, petrified, he checked his seat belt and discovered that without consciously thinking about it he must have pulled it as tight as possible the moment he first heard the scraping noises.

After watching the blaze for a second or two, he again put his head down and, leaning forward as far as he could, resumed his "brace for impact" position.

2340h44. Suddenly the two engines beneath the left wing also started to scrape their way into the earth. In the last second, further pieces, now more substantial, had begun falling off the Clipper. Immediately after the front undercarriage touched, the radome (the black cover on the very nose of the airplane which conceals the scanning weather radar transceiver) fell away and was crushed by the weight of the skidding machine. The door on the forward wheelwell was ripped off when the landing leg collapsed.

Sixty-five feet further on, a sixteen-foot-long section of the right wing, including the outboard flap, was detached.

The front undercarriage steering cylinders were torn from their mounts. Then the left engines dug into the lush topsoil.

The pilots looked up to see onrushing vegetation flapping, tapping lightly against the windshield as though trying to tell them something before it was mowed down, dismembered by the fast-moving airplane. They were stunned, hypnotized by the dim scene confronting them.

The power plants may have started to wind up after Phillips opened the throttles, but now they abruptly stopped, jammed internally with dirt and minced plant life. However, the fuselage was still completely intact, as far as anyone could determine.

As the plane continued to jolt along the uneven ground, Roger Cann grasped his safety belt. He decided there was no point in undoing it until the slide had actually ceased. He checked the positions of the emergency exits, and made up his mind that when the Clipper did come to a halt, he and his wife would get out as quickly as possible. As the bumping and shaking continued, he called over to Heather: "Get ready to release your seat belt," and grasped the buckle on his with his left hand, readying his right index finger on the metal release tongue. He saw out of the corner of his eye that his wife was doing the same thing.

Even though he had taken all these precautions, he was still perplexed. Was this a normal situation, or wasn't it?

A second or so after she had lowered her hands to her belt, Heather Cann, preparing for a pile-up, leaned forward and grabbed the back of the seat in front of her. Again she glanced out of the window. Now she saw that the intensity of the flames had increased. The machine was still shuddering and jerking as it scraped its way along the densely overgrown jungle floor.

Dick Smith had also made up his mind that somehow he would get out of this thing—*alive*. He knew he was prepared. His seat belt was tight. He was securely propped in one of the positions recommended by the airlines themselves. The sounds entering his head were the scraping of the lower fuselage as it grazed along the jungle and the regular thumping and thudding as the wings, like butter knives, sliced their way through the foliage.

2340h45. Suddenly Clipper 806 transmitted a bone-shuddering ja to those on board as a section of its right outboard wing crashed into a rigidly constructed three-foot-high lava fence. Microseconds later, the starboard portion of its nose ploughed through the solid wall, flattening it.

The pilots were strapped in their seats with a three-way belt system. To the normal waist band, similar to those used by the passengers, was looped two harnesses, one for either shoulder. Also, two leg straps were fastened to the front of the seat and clipped on to buckles near the armrests. Jim Phillips, like everyone else, had been expecting a routine landing. He had neglected to affix his leg straps, though the only partly secured waist and shoulder restraints managed to hold him roughly in place.

But as they exploded through the volcanic rocks, the right side of the cockpit next to him was ripped open and almost completely crushed. In that second of impact, both Phillips' legs, both his arms, and a number of fingers—mainly those on his right hand—were broken. He was suddenly helpless, confined loosely to his seat by the dangling shackles.

Dick Smith's subconscious was, by now, working overtime. He was sure that this was an accident situation. No, not "he was sure," he knew it was. But where were they? What were they crashing on? Or into? It felt and sounded as though they were sliding along land, but next could be water. *"Well, that won't make much difference,"* the conversation went on in his head. *"You know how to swim through oil slicks, amid floating debris, and in brine that is covered with fuel, Smith. You also know how to evacuate from sinking objects."* *"Yes,"* he replied to himself, *"that's right."*

Then came this tremendous jolt. The airplane had met

something head on. The collision jerked Smith's upper torso forward and his head banged into the rear of the seat facing him, gashing the brow above his right eye. Although it was bleeding heavily, at that moment he was unaware he had been injured.

As the traction generated by the encounter with the fence subsided, Smith thought the Clipper had come to a stop. Momentarily, everything seemed calm, serene, peaceful.

In the collision with the wall, Roger Cann, like nearly everyone else, was thrown forward as far as his seat belt allowed him to be. After the jolt, he instinctively tried to tighten the webbing further. But he still had the impression that the forces were not strong enough to cause physical injuries. He had not struck anything.

Heather Cann had been hurled upwards and out of her belt. She suddenly realized she was dangling, leaning over the top of the back-rest in front of her, with the restraint down around her knees. She had not hit anyone, so she thought, *"Well, there seems to be nobody in this seat."*

2340h46. The airplane continued jolting, less severely now, as it sliced through the jungle, hacking down small trees, palm groves, vines, and vegetation, and spraying dismembered foliage in all directions. As it flattened the lava fence, the starboard outer aileron—one of the flight control panels on the rear of the wing—came loose, together with a section of the keel beam and the jackscrew that operated the right wing's flap.

Once past the wall, the inboard high-speed aileron detached from the right airfoil, and the nose landing gear, which had collapsed back down the crash path and had been dragged along ever since, was torn away.

2340h47. The rate at which the fuselage was sliding through the vegetation had greatly decreased, though whenever it struck something more solid than the normal lushness, other sections of the still fairly undamaged body were wrenched off.

Now the floor of the jungle was flat and even. Many of

the passengers had the feeling that the machine had actually stopped. A number of them leaped out of their seats and began heading for the doors. The first signs of panic, frustration, bewilderment became apparent.

Meanwhile, the strut holding the inboard engine to the right wing was dislodged, then a number of the operational workings of the left-main landing gear and a twenty-six-foot-portion of the port wing joined it, all in two and a half seconds. At the same time the flames, burning fiercely around the remaining stubs of both airfoils, had spread to the flattened jungle itself. They were being fed by some of the 8,375 gallons of kerosene that had still been on board. Large holes had been ripped in the tanks and a drenching carpet of the liquid was being laid, sprayed behind, as the plane continued its slide. The kerosene was being ignited by the fires coming from somewhere underneath the fuselage.

2340h48. Suddenly, an extra large fuel spillage occurred. It was instantly kindled. A huge ball of flames went streaming up into the midnight air, momentarily lighting the entire surrounding area like a luminous pink sunset.

Back in the terminal's coffee shop the electricity flickered. Someone near James Prendergast's table commented that it was more than likely a warning that the power would be going off altogether. The air-taxi mechanic, Cary Wade, peered through the open window space to see whether there was a thunderstorm in the vicinity. All he saw was a brilliant glow coming from the far end of the airport. He stared at it intently. He could not tell what it was. Its outline was blurred by the torrential rain still falling nearby. He assumed that whatever it was, it had caused the building's lights to flash.

Prendergast did not see the fireworks outside. His attention was on the next table; he was trying to identify the person who had made the remark about the possible blackout.

George A. Wray, the owner and manager of South Pacific Island Airways, the air-taxi company for which Wade worked, was driving toward the airport when he, too, saw the sky light up with an intense glow that appeared to be originating at the far end of the runway. It remained for about three seconds. He couldn't tell what it was either. In the last few minutes the downpour had lashed his windshield so heavily that he had been forced to slow to a crawl. He had been concentrating on the road ahead. However, he couldn't have missed the shimmering reflection of the ball of flames against the wet surface of the pavement, even though it had originated some distance away.

2340h49. Dramatically, and with an enormous impact, the Clipper ground to a surprise stop. A moment before, it had slid over and crushed the runway's final lead-in navigational beacon, known as the "middle marker." This transmitter was stationed just 3,090 feet before the start of the concrete.

The aircraft came to rest poised over a small gully. The underside of the central fuselage was held aloft by the remains of both wings and the tail section; they were resting on mounds of earth on the sides of the tiny ravine.

Outside, the entire central section of the passenger cabin was enveloped in roaring flames. Back along the crash path, as the fuel that had been spilled was consumed, and as the tropical rainstorm reached the area, the fires slowly went out.

6

When the aircraft finally ended its wild, yet straight, slide through the jungle, Dick Smith remained for a moment in his seat trying to decide what to do. In a state of shock, he looked about him, half taking in the situation in the cabin. He became aware of a lot of activity—many people were out of their places, moving both forward and aft. His subconscious again began talking. *"Now look here Smith,"* it whispered, *"sit tight. Take it easy. Hurry up, but take your time."* For a few seconds he stayed where he was, simply watching.

Without having finally decided what to do he released his seat belt and stood up. He then bent down to look out of the windows. The fire was still there—now more furious than before and radiating its heat inside. He stooped further, attempting to see how high the flames were jumping. They were going up well above the top of the cabin. He again stood erect, then half-turned toward the center of the fuselage. The area around him was empty of people. He paused momentarily, wondering where on earth the young Australian girl had gone. He then noticed that the Samoan children who had been about three rows in front of him, as well as the two American mountaineers, had also departed. *"Jesus, I'd better get out of here,"* he told himself.

He quickly turned toward the windows and partly opened the emergency exit. The moment the top of the panel had released, he was struck in the face by a sheet of flames. With all his strength he jammed the section in its position and re-locked it. Dazed, he sat down.

When the plane impacted and stopped, Heather Cann had been thrown back into her seat, the loose belt still clinging around her knees. She heard her husband Roger tell her to unfasten the harness. This had been their original plan, but now she could neither laugh nor cry. She could only sit there, stupefied.

Meanwhile, Roger, who had been concentrating on his own problems, undid his seat belt, jumped up, and pushed his way into the central corridor past Charles Culbertson, next to him in 16-C.

Impact + 1 second. Pausing to take in the situation, Roger saw Michael Merrill, in 17-A, stand and move forward into the space where 18-A would have been had it been installed, and start struggling with the rear emergency exit panel situated there. He also noticed that many other passengers were already on their feet, some working their ways toward the front and rear doors. He could see all this clearly because, even though none of the electric lights inside the machine were on (the emergency system was not operating and the aircraft's power had flickered and failed a few seconds earlier) there was ample luminosity streaming through the windows from the fires that encircled them outside.

Impact + 2 seconds. Roger noticed that Merrill was tussling with the exit panel. Possibly, Roger thought, all of the kit bags and paraphernalia that had been stored in that space before the flight was getting in his way. He stood watching for a few moments, trying to decide in which direction he and his wife should go. Then he moved forward to the end of the row in which Merrill was so desperately attempting to open the window.

Charles Culbertson remained in his seat, stunned.

Heather Cann by this time was having trouble unbuckling her belt, now back around her waist.

Dick Smith, four rows ahead of the Canns on the other side of the cabin, was still recovering from the shock of being hit in the face and hands by the flames. Now he regained his presence of mind and stood up once more. People were still rushing past, hurrying to the fore and aft exits. He glanced from one group to the other, trying to decide where to go.

Impact + 3 seconds. Realizing it would be stupid to fight against the main traffic stream heading toward the front of the plane, Smith moved past the aisle seat where the young Australian woman had been. The important thing was to get into that passageway.

Merrill—at last—had partly unlocked the rear-left emergency window and Roger Cann watched him as he wrenched the entire panel straight inside. A sheet of flames or fumes gushed through the opening. Merrill staggered backwards.

Roger was standing in the aisle next to 18-C. Noticing what had happened, he glanced through the portholes on either side of the fuselage and all he could see was flames. After having concentrated so completely on how he and his wife would leave the plane, he now realized for the first time that they might not be able to. They were in a tube surrounded by fire.

But there seemed to be no major panic from the others around him, just the steady movement forward and aft. Some of the people were softly murmuring to themselves. All could certainly hear the sizzling of the encircling inferno.

Impact + 4 seconds. The air was suddenly pierced by a scream from a woman somewhere up front—a cry more of terror than of distress. It lasted for hardly a second, reverberating around the cabin. Then the main background noises returned: the low-pitched moans, the rustling of the moving passengers, the spitting of the outside fire.

Dick Smith was now poised in the central passageway, listening. The area around him had been completely vacated. Up front, he could hear a great deal of noise and commotion, verging on outright panic.

Roger, motionless at the end of the row of seats, was waiting to see what Merrill would do next. He did not realize that his wife was still struggling to release her safety belt.

Impact + 5 seconds. Back in the terminal's coffee shop, Jay McLean, the air-taxi pilot, looked worried. He had leaned over to his mechanic, Cary Wade, and whispered that he had seen a flash which had seemed extraordinarily like "a fire on finals"—aviation jargon for an aircraft explosion in the latter stages of a landing.

James Prendergast, seeing the anguished expression on McLean's face, gestured for him to follow. He headed for the coffee shop's door. Wade remained seated, disturbed by McLean's comment.

The two Samoans in the fire truck parked alongside the runway were still awaiting the arrival of Clipper 806. With the rain pouring down, they hoped the aircraft was not too far away.

Frank D. Bateman, the air traffic controller, was also waiting. Sitting at his desk looking blankly out the small distant window, he was startled to see a hazy red glow that apparently was coming from somewhere down the far end of the runway. He did not have direct communication with the truck stationed beside the landing strip so he decided to contact the officer on duty in the airport's emergencies station. He switched on his microphone and, in a persistent tone of voice, said "Fire Two. Fire . . . Fire Two. What happened?"

Impact + 6 seconds. Heather Cann had finally released her seat belt. She sat still, staring fixedly at the two loose ends of the webbing, one in each hand.

Charles Culbertson was standing in the aisle next to his seat, trying to decide in which direction to escape.

Dick Smith had decided. However he had taken only a step or two forward when he was abruptly stopped by a loud, authoritative male voice, calling from somewhere up front "For God's sake. WILL YOU QUIET DOWN!" After that, even the soft moaning seemed to hush.

Impact + 7 seconds. Smith stood motionless once more, possibly as a reaction to the command. Suddenly he noticed something encircling him, stifling him. It was either smoke or a heavy, noxious cloud of fumes. He didn't know whether he was knocked down or instinctively dropped, but the next moment he found himself on his knees in the aisle.

His mind raced back to earlier experiences at his Swim Gym when he had accidentally been exposed to chlorine gas. The feeling was like being smothered in a vat of thick molasses, but quicker. In a single part-breath, whatever it was had grabbed him around the throat and seemed to be trying to strangle him.

Smith stayed in his kneeling position. Why was he down there? he asked himself. It could have been because of his nearly thirty years of military service, during which he had undergone a great deal of "escape and evasion" training. It had taught him that "good air"—if, indeed, there was any—would be near the ground. His earlier episodes with the chlorine gas made him immediately exhale and stop breathing the moment he had taken just a part of one lungful of the toxic substance. He knew that holding his breath would not be too difficult. In swimming he had been able to stay underwater for up to three or three and a half minutes. However, he also knew he would have to get out of the plane quickly, before his "breath time" was up.

Panic now seized him. He became afraid of going on, yet he was more afraid of dying. He knew all he had to do was to take one whiff, and that would be it. *"That's the best,* easiest way out," he told himself. "*It will save you from having to face a slower, more painful death in the fire.*"

Then, gradually, he got hold of himself. He tried to remember the way things had looked just before he fell to his knees. There had been no electric lights on in the cabin, or if there had been, they were completely blocked out by the dense smoke. Immediately in front of him, he remembered, there had been a line of waiting, jostling people, which, he assumed, stretched all the way to the forward passenger door. He could not recall having seen any of the stewardesses. Neither had he heard any of them giving orders. He continued holding his breath, keeping his eyes tightly closed.

Impact + 8 seconds. Up in the cockpit, it had taken Captain Petersen and his crew a very short time to close down the main controls and disconnect the circuit-breakers. Then Petersen looked toward Jim Phillips. He was slumped in the co-pilot's seat, trying without success to remove the loose belts. It was hopeless; both his arms and legs and many of his fingers were broken.

The captain stretched over the central console and released the harness for him. Phillips wriggled and struggled, and finally fell out of the large hole just to his right, the one made in the collision with the lava wall.

Petersen twisted around. The cockpit door was open. His flight engineer had gone. He had turned just in time to see the first officer's back as he, too, made his way into the forward area of the passenger compartment, evidently to assist with an escape.

Impact + 9 seconds. Halfway down the cabin, Merrill dropped the exit slab and hurled himself through the opening into the roaring inferno, hoping to land on top of what he presumed was the fire-hidden wing.

Roger Cann, seeing Merrill's feet disappear, decided that he, too, would escape through there. He moved toward the window. The panel Merrill had thrown aside was now lying angularly across the opening. Luggage was still piled up behind the partially blocked exit. What was more, he had just noticed that the two seats ahead of him in row 19 had broken backwards, further impeding his way. He realized that he would have to reset the seat backs before he could remove the kit bags and the panel; *then* he could make his getaway.

Although there had been no instructions from the crew, he knew what he was going to do. He was sure his wife also knew what to do. It flashed through his mind that if this was a real air crash, it was "*a piece of cake.*" As he started to push the things out of his path he felt certain most of the passengers would survive.

Impact + 10 seconds. Heather Cann's brain at last began to clear. It was no good simply sitting there, she had better move. But first she bent forward to retrieve her handbag. It was stuck under the seat in front of her. It should not have been. She had simply placed it under there prior to the landing, and now it was stuck.

Impact + 11 seconds. Even after giving it a few solid tugs, she could not free the damned thing. She decided to leave it where it was and stand up. Once on her feet, she looked around. Her husband had disappeared. Charles Culbertson had vanished. She made her way into the aisle.

Impact + 12 seconds. Dick Smith had regained his composure. The quiet moaning up front had stopped.

He decided that the best survival route lay behind him. He turned on his knees and, facing the aircraft's tail, cautiously opened his eyes. The first thing he noticed was that the general illumination inside the cabin was "flashing." It was not the electric lights. It was "constant undulating reflections," bright one second, then dim like the glow from a fire.

Impact + 13 seconds. Roger Cann at last had pushed the seat backs up into their erect positions. The next thing he had to do was to move the luggage, then the window panel from the area.

As he began the job, he thought his wife was directly

behind him, possibly in the passageway, waiting for the path to be cleared. His attention was briefly attracted to a shadowy form outside in the middle of the flames that hid the wing from view. Somehow he imagined that *that* was Heather, making her getaway through the fire. Now he worked harder than ever, thinking that he only had himself to worry about.

Impact + 14 seconds. As Heather Cann arrived at the end of row 16, a woman brushed past her heading toward the rear of the plane. She was somewhat bigger than Mrs. Cann herself, and appeared to be European. Heather did not see where she had come from or where she went, but it occurred to her that the woman must have been, of the two of them, first out of her seat.

Impact + 15 seconds. Dick Smith, still holding his breath, had crawled back until he was next to row 20. He was searching for the forward-left over-wing window, and saw the removable panel rather than the word *Exit*—if, indeed, the sign was even there any longer, covering the retention handle. He decided to scramble along the floor between the seats and get the egress unlocked.

Impact + 16 seconds. Meanwhile, Fiaoo Masina and Laalaai Niko patiently sat in the truck beside the Pago Pago runway, waiting.

Impact + 17 seconds. Smith arrived at the emergency window and looked up at it from his kneeling position. He still had not breathed, and had a choking sensation in his nose and throat from that slight inhalation of fumes or smoke. Never in his life had he experienced anything like it. He said to himself, *"No, Smith. Not you. Not this way."*

With these thoughts running through his mind, he glanced back toward the exit hatch on the opposite side of the cabin, trying to select the safest route. One looked as bad as the other. But he told himself, *"Now you have two ways out."* He chose the closest.

As he did so, he realized that the whimpering sounds had stopped. His experience with the fumes made him suspect that the passengers up front would live only as long as they could manage to hold their breaths. For the average person, he knew, that was only a matter of, say, a quarter of a minute.

Impact + 18 seconds. Two rows behind Smith, Roger Cann was frenziedly tearing away the last of the obstacles that had blocked his path. He was so intent on what he was doing that he was conscious of neither the acrid smoke nor the other people.

He removed the final kit bag at last and, stooping to avoid hitting his head on the hat racks, grabbed the panel. Since there seemed to be no convenient place to put it, he tipped it on end and hurled it through the opening as far out as he could.

Impact + 19 seconds. With the exit clear, Roger leaped head-first through it, twisting his body in midair so he would land ready to roll through the fire.

Heather Cann, beside row 18, saw her husband's back disappear into the inferno. As she moved toward the open space, unexpectedly, coming from her left, Charles Culbertson stepped between her and the escape route.

Impact + 20 seconds. Meanwhile, two rows in front, Smith had climbed to his feet next to the still-closed forward window. He was determined to take no chances with the outside flames this time. With his head tilted and one eye tightly closed, he grasped the release handle in his outstretched hand. After a short struggle, it unfastened. He slowly pulled it inwards. By now he had recalled his military training, which had taught him how to open escapeways in the vicinity of fire.

When he had fully released it, he turned to throw the heavy frame out into the inferno. Glancing down as he did, he saw something on seat 20-C. He was not sure what it was. It could have been either one of the Pan Am pillows which had been placed at the disposal of the passengers in the early part of the flight, or it could have been a sweater left behind by the young Samoan girl who had been sitting there. He cast the window panel outside and picked up the object that had caught his attention.

With a final quick inspection of the exit opening, Smith covered his face with the pillow or sweater and dived through, releasing whatever it was he was holding so he would have his hands free when he landed. It fell away and was incinerated.

Impact + 21 seconds. Back in the air traffic control office, Frank Bateman had been anticipating a reply to his urgent call to the duty officer in the fire and rescue building. None arrived. Now he decided to try and make radio contact with the Boeing 707's flight crew. "Clipper Eight Oh Six. Pago Approach."

Again, he waited.

At the same time, Charles Culbertson, who had been poised to get out the rear-left emergency window, the one through which both Merrill and Roger Cann had escaped, felt a movement behind him. It was Heather Cann. Gentlemanlike, Culbertson stepped aside, and Heather, grasping the opportunity, shut her eyes tightly, took a deep breath, and half-dived, half-jumped through the opening. When she had inhaled, however, she had pulled in a lungful of acrid fumes. They had a distinctive, pungent smell that was to stay in her nose and throat for some time to come, bringing on bouts of uncontrollable coughing.

Impact + 22 seconds. Roger Cann had landed in the flames on his side, as he had planned. He immediately started rolling over and over. He had come down onto something soft. He thought it might be mowed foliage—wet, possibly drenched with kerosene. He was afraid to open his eyes to see what it was, but it didn't feel like the aircraft's wing. Besides, it appeared to be the source of the fire itself. He didn't know whether the wing was still in position. He thought it must have been ripped off and perhaps was somewhere back along the crash path.

Impact + 23 seconds. Smith had come down in the heart of the fire, but in a trampoline position on his hands and knees. He had decided on this stance because he had expected to land on a slippery metallic surface.

He pushed himself to his feet using his right hand. Once erect, he cautiously opened one eye and thought he saw what he later described as "a big long aluminum highway to freedom." Closing the eyelid, he blindly headed away in the direction, he presumed, of the wing tip. When he imagined he was halfway along the airfoil, he again slightly, briefly opened one eye and glimpsed, there in the flames, what looked like the silhouette of a man, in front and marginally to the left of his position—somewhere, he thought, near the wing's trailing edge. Between blinks, Smith noticed this man—who had his back to him and his clothes burning like a torch—suddenly throw his hands above his head and fall. Still running, Smith saw him get to his feet once more. Then the chap's legs became tangled and he slid backwards, calling out "Oh my God. Please help," as he dropped from sight into the fire below. Smith, completely surrounded by flames, continued running, only momentarily distracted by the agonized cry.

Impact + 24 seconds. Roger Cann pressed on, rolling from side to side until he reached an area where the flames were noticeably less intense. Then he jumped up and started to crawl and run sightlessly over the burning debris that covered what he presumed must be the wing. (It was most likely the earth mound at the side of the little ravine on which the port wing stub rested.)

His wife was not too far behind him. Heather had emerged on her feet and, eyes firmly closed, immediately began to rush away from the melting fuselage. Suddenly, she stumbled into a furrow. As she was falling, she opened her eyes slightly and thought she saw the outlines of a couple nearby in the fire. The man appeared to be half-dragging the woman. They were just about level with Heather but slightly nearer the airplane's tail.

Heather landed on her left knee. To save herself from falling full length, she put both hands out in front of her.

As she sank, instead of metal she felt soft foliage. It seemed to be floating on top of mud, or slush, or something that was very, very wet.

Impact + 25 seconds. Heather knew she must quickly get up again and keep moving. She regained a semisteady footing and began to run.

By this time, Roger was on his feet too. After blindly racing ten strides or so, he thought to himself, *"Well, if I'm not out of these flames pretty quickly, I'm going to be dead."* He had not been conscious of passing the tip of the wing. His whole escape route seemed to be over the same type of surface, mowed-down vegetation. The fire continued, the flames leaping high overhead. He could feel them burning the skin off his face and hands.

Impact + 26 seconds. The airport's traffic controller, Frank Bateman, still awaiting replies to his earlier calls, was puzzled. Once more he tried to contact the brigade headquarters: "Fire Two. Fire Two."

He waited.

By now, Dick Smith had crawled through the flames, arriving at the trailing edge of the wing somewhere beyond the outer engine. With a mighty frog's leap he hurled himself off it and collided heavily with something solid. *"My God,"* he thought, *"we must be in really rugged terrain."* He had struck the lava bank on the side of the ravine. The shock stopped him for a moment. He thanked God he was still alive, even though he was, at this time, still surrounded by the blaze. He remembered an event in the past when he had seen a military plane afire. It had been coming in for a landing at a base in Germany just after the end of the Second World War. While he was watching, it had suddenly exploded. He must keep going, he knew, and fast.

Running up the bank and over a small level plateau, he unexpectedly came to the edge of the inferno. Without pausing, he pushed his way into and through the cool, rain-drenched underbrush.

Impact + 27 seconds. When he was about ten strides from the limit of the flames, Smith decided he had gone far enough. He sat down and turned to face the horror from which he had just escaped. He suspected that what remained might explode at any moment, disintegrating completely in an enormous ball of fire.

The memories of other aircraft accidents he had witnessed began returning to him now. He had been a Special Services physical education officer with the U.S.A.F. in Britain during World War II, assigned to the Fifty-fifth Fighter Group. He remembered the many occasions he had seen his unit's planes crash on take-off or on landing, a number of them being enveloped in balls of flame as he watched. There was one time, he recalled, before he was sent to Europe, when he was officer of the day at the Troop Carrier Command in Indianapolis. A C-47 from another area was in trouble and had been given permission to land on his base. As he watched it come in for a touch-down, it suddenly dropped to the concrete, bounced, and running out of control, came to a stop right in the middle of a squadron of cub aircraft. Then it exploded, taking a number of the others up with it.

Then the most hair-raising experience of all came back to him. It had happened a few years earlier, when he was with some friends in a Navion four-seater and they were making a landing approach to Kansas City Municipal Airport. As the pilot was lining up with the runway, unexpectedly, an oil-line broke; the dark, heavy liquid spurted from the rupture; the engine began to splutter, liable to seize at any moment. The pilot already had his hands full simply trying to fight strong crosswinds and keep them aligned with the runway. Smith remembered being convinced that *that* episode would be the last of his life. It was not. Somehow they made it down safely. The plane was afire when it came to a halt. Smith scrambled out of the cabin and ran full speed away from it, just in case it did blow up.

Now, as he rested, he could see that the blaze on the far side of Clipper 806 was intense. He had never seen such a furnace before. Its source seemed to be underneath the airframe itself, with, every so often, small bursts leaping high into the sky, momentarily lighting up the entire area. All the while he could hear a kind of spewing or gushing noise that made him decide with a befuddled irrationality that the wreckage was not going to explode after all. His mind went back over those moments when the aircraft was sliding along its crash path. He recalled hearing the undercarriages being ripped off. He remembered the terrible scraping, scratching, tearing noises. He reasoned that the fuel and hydraulic lines, likewise, must have been torn loose; possibly the gurglings he was hearing now were the last of the combustibles draining from the tanks. Anyway, he thought, if it did explode he was far enough away to have a chance of surviving, of retreating as quickly as possible still further into the dense, cool jungle.

Now and then as he watched, the fire on the far side of the wreck would billow up wildly. This seemed strange to him since the plane itself appeared to be flat on the ground. In the area he was looking at the flames would flare for a moment and then die down completely, whereas around the middle of the fuselage, and especially on the side opposite him, they were much more ferocious. They were also very violent next to the cabin on the side nearest to him. The worst of it seemed to be concentrated in the vicinity of the wing—from where he had just escaped.

As he studied and restudied the nightmare, it dawned on him that even though there were a lot of low-level flames, the front of the aircraft was intact. The fuselage itself looked complete. The doors were shut. The metalwork seemed to be undented and undamaged. However, the tail was a little twisted, slightly "torqued" in his direction.

Because of the relative wholeness of the airframe, Smith somehow surmised that the burning nose was facing toward the airport. He considered, for a moment, walking along the line it was pointing. But he soon dismissed this idea because he didn't know how far away it was. He stood up. He could see no lights, no paths or roads. He could hear no sounds—nothing that might indicate the direction to an inhabited area. All he knew was that the Clipper had been in a landing pattern; therefore, the strip must be somewhere close by. But where, and how far, were mysteries.

He again looked toward the burning wreckage. Staring blankly at it, he felt certain that by now all the other people on board must have escaped. He sat down and continued watching the blaze, then suddenly realized it was raining; the water was literally pouring out of the midnight sky. He remembered that right after he had opened the emergency window he had thought it was showering too—but he hadn't, like now, been soaked to the skin. It was a good thing, he considered. It would stop the fires from spreading. And it certainly would comfort those who had been burned in the getaway.

"*Those people . . .* " he thought. He remembered seeing the man fall off the back of the wing. He pictured the whole episode as though it were a slow-motion television replay. Even now he could hear in his mind that cry of agony, *"Oh my God. Please help."* He decided to retrace his steps, find the others and offer assistance.

He returned to the edge of the flames near where the jagged, broken left wing stub was resting on the lava mound at the fringe of the tiny ravine. Just as he arrived, Merrill was getting to his feet. Smith went over to him. For a second or so he stood there, looking. Most of Merrill's clothes had been burned off. What remained seemed to be hanging together by threads. It was not afire. Smith could see that where his skin was exposed, it was white and very puffy. He was brought back to reality when Merrill said to him, "Help. My wife. My wife. Oh my God, my wife," and pointed to the ground. She was lying on her side in the rain-soaked foliage, practically at their feet. Smith reached down, put his arms under hers and gently lifted her to a standing position.

Just then another voice was heard over the sounds of the crackling fire and the patter of the rain: "Can I go with you?"

It seemed to be coming from the same spot where Mrs. Merrill had been lying. In fact Smith now realized that Mrs. Merrill had fallen on top of this other woman.

Smith offered to help her get up.

"No. Not my hands," she begged, evidently in agony.

Smith looked at her. She was very seriously injured. She was much younger than Mrs. Merrill. All her clothing had been burned off; she was practically naked. Somehow he helped her to her feet.

Then he took a closer look. She was a terrible mess. Her blouse appeared as though it had exploded off her body. Her panty-hose could be seen, in the flickering of the nearby inferno, to have melted and welded onto her skin. Her hands were in desperate shape—burned raw and bloated by the unrelenting flames.

He looked toward Mrs. Merrill. She had her back to him. From what he could see, her clothes appeared to be relatively untouched by the fire. She was wearing a light dress, scorched only slightly here and there.

Suddenly, for some unknown reason, panic broke out in their group. The women began to yell and scream. Smith, in a loud, commanding voice, ordered them to stop. "Follow me," he insisted. "Come on, I know the way out of here." And he started to move away from the heat and into the cool of the jungle.

A moment or two later, the others decided to follow.

Mrs. Merrill appeared to be having difficulty walking, so Smith picked her up in his arms. Behind him, hobbling slowly, came her husband. He was very badly burned. Heather Cann was a few strides further back.

They had moved only a short distance when a voice, coming from the undergrowth, called out "Heather . . . Heather." It was Roger Cann searching for his wife.

After about twenty agonizing paces, choosing their way more by feel than by sight—the only light came from the blazing wreckage they had just left—they arrived at a small clearing. Smith gently leaned Mrs. Merrill against a tree trunk. Even though most of her clothing was intact, her face looked so ghastly he decided she must have suffered some terrible head injuries.

While the rest of them were slowly moving into the little clearing, Roger Cann, at the rear of the file, also realized it was raining. He would not have called it heavy; it was more of a medium shower by New Zealand standards. As he waited for his turn to enter the glade he noticed that both his hands were burned. He could then feel the same painful condition on his face.

Impact + 51 seconds. James Prendergast, trying to keep up with air-taxi pilot Jay McLean, worked his way through the crowded coffee shop, sidestepping the chairs and tables which everyone had moved back away from the window spaces as the downpour increased. Glancing out of one of the openings as he went, he saw the rotating beacon on top of the fire truck parked next to the windsock, near the runway. For some reason, his mind connected the flashing hazard indicator with the tenseness on McLean's face.

Once outside, they moved down the corridor to a balcony. There, McLean stopped. Breathlessly he pointed to a red glow coming from somewhere not far beyond the end of the landing strip. The two men looked at each other. The same thought went through their minds.

They raced to a nearby telephone and tried to call traffic control. The phone was ringing, but Frank Bateman ignored it—he was busy at that moment. He was listening to a radio conversation that had recently begun between the Pago Pago Port Management and the duty officer in the airport's emergencies station:

"Fire Two. Port Central."

"Roger. Go ahead, Port Central."

"Hey . . . " The speaker at the other end mumbled on for several seconds. It was impossible to make out what was being said.

When they had finished, the fire officer simply replied "Roger."

Bateman was worried because, now that the rain had slightly eased, he could see the flickering glow at the far end of the field more clearly. He still could not make out what it was. He recalled that both his earlier attempts at raising "Fire Two" and the due-in Pan American 707 had failed. After the radio exchange had ended, he waited a second or so and then depressed his microphone key: "Fire Two. What Happened? Port Central. Anybody. Do you read?"

Impact + 70 seconds. Prendergast and McLean, having no luck in trying to reach the communications center by phone, decided to make tracks for the fire headquarters. They hurried toward the parking lot and Jay McLean's van.

Meanwhile, the bowser (R-1) remained peacefully at its station halfway along the Pago Pago runway, still awaiting the arrival of Clipper 806. Its red rotating dome light continued to flash, the glow penetrating yet indistinct, blurred by the seemingly never-ending cloudburst.

Not long after the five survivors had reached the tiny clearing and while they were trying to decide what to do next, they heard what they thought was approaching footsteps—a spasmodic rustling sound in the leaves and bushes. It continued for a moment, then stopped. They dismissed it as being the effects of the rain. Then the noises started up again and grew louder. Charles Culbertson, escapee number six, entered the area. He had been attracted to the group by their voices.

Culbertson's arrival made Smith and Roger Cann wonder whether there could be any other survivors nearby. Cann ventured back along the partly worn path toward the still-ferocious fires that were eating into the airframe. Because he too thought the plane might explode, he kept a respectable distance. He walked around it in a clockwise direction, heading toward the nose. He could neither hear nor see anyone. He stood motionless for a moment and thought *"Well, what are they all doing? There has to be some more people somewhere. They couldn't have all died in that one. They must have escaped out of the other side, or even gone farther into the jungle. They've got to be somewhere."*

He turned and started back the way he had come. It then occurred to him that his wife would get anxious if he stayed away too long.

Cann had no great difficulty in finding the others—most of them were suffering from the effects of their burns and were moaning aloud. There was no panic now among the members of the group. Smith was quietly talking to them, telling them that rescuers were on the way and soon they would be out of there. However, other than by Smith, very few words were spoken. The injuries were too agonizing; the pain too intense.

Impact + 84 seconds. Frank Bateman, the controller, had still received no reply to his agitated calls for information. He decided to try once again: "Fire Two. Fire Two."

This time, the silence was broken.

"Roger, Pago."

"OK. What happened? Please tell me."

"We've just received a call from the other ones standing by for the aircraft, that there's a fire down the end of Runway Zero Four."

Moments earlier, Fiaoo Masina, the driver of R-1, had radioed to his station that he and his assistant could see something burning toward the western end of the airstrip. They reported that the Pan Am aircraft had not landed and they would stay where they were until it had done so before checking what it was. The trouble, he said, seemed to be coming from a position just a short distance outside the airport's official perimeter, between the runway and Logotala Hill.

Robert S. Sievers was Pan American's local agent in Pago Pago. He collected the aircraft's pre-cabled maintenance reports and the loading documentation and, looking at his watch, noticed that the flight was due in at any moment. He hurried toward the assigned parking ramp. The tropical shower had intensified somewhat. The rain was again pelting down outside, accompanied now by very blustery winds. He was relieved to see that he had arrived at the gate before the Clipper. He stood near the partly opened boarding door and waited.

By this time, Frank Bateman was really frustrated. He could not seem to get any logical explanation for the flickering at the far end of the runway. He picked up the telephone and called his boss, Carl Schumacher, at home, telling him that there was *something* at the incoming end of the airfield that was glowing in the night sky. He reported he could get no sensible answers from the fire station; neither could he raise the Pan Am pilots on his radio.

Schumacher listened to Bateman's story, then phoned Tommy Bracken, the Federal Aviation Administration's resident director and section manager for the islands. Schumacher told him that Bateman believed an incoming P.A.A. flight had "bought the farm" on Logotala Hill. It took Bracken no longer than sixty seconds to get dressed and, while doing so, he instructed his wife, June, to call the hospitals and alert them for a possible accident. He told her not to give them any details because, actually, he had none to report.

Impact + 120 seconds. Traffic director Bateman was fairly certain in his own mind that the red glow had some connection with the 707. He picked up his headset that put him in direct contact with the area control office in Fiji. "Nadi. This is Pago."

There was a slight pause.

"Nadi," came the reply.

"OK. This is Pago. I think I've got one. My Clipper Eight Oh Six, I think, 'went in' right here on approach."

"You're kidding me."

"No I'm not," Bateman insisted. "I don't know yet. Just a big ball of flames at the end of the runway, and I can't get too much out of this crash crew yet. So you might alert your authorities."

"OK."
"OK. Thank you."
Prendergast and McLean had, by now, arrived at the fire station. When they entered, the officer on duty, Sainila Fanene, was involved in a long and animated radio conversation in Samoan. He was being told by Fiaoo Masina that they could still see the something that was burning outside the airport's boundary in the direction of the incoming aircraft.

Yes, the truck driver stated, the flames were visible from their position. But what he had decided to do was to wait until the due-to-land 707 had touched down, *then* he and his assistant would go and attend to the matter.

Impact + 250 seconds. The general atmosphere in the air terminal was one of serenity. Only a handful were puzzled by what might be causing that blurred glow a few thousand feet away; most had not even noticed it. The tropical rainstorm raged on in all its intensity, joined periodically by strong gusts of wind. Cary Wade drank another cup of coffee in the cafeteria. Traffic manager Bateman was uncertain and anxious, wondering whether his eyesight had been playing an untimely practical joke on him. From what he had been able to determine, the fire station seemed quiet; the operations office seemed quiet; and the words of the Nadi area controller, "*You're kidding me*," went around and around in his mind. Possibly he had been wrong all along. Possibly it was something else. Then if it was, he suddenly realized, he should be able to talk with the incoming pilots. Again he pushed his microphone key: "Clipper Eight Oh Six. Clipper Eight Oh Six. Do you read?"

He breathlessly waited a few seconds.

Still no response.

"Fire Two or Port Central. Fire Two or Port Central. This is Pago. Can you give me any information?"

The brigade's duty officer answered his call: "I've just had another report form our R-1 (Rescue One) that a fire has occurred to . . . " The voice trailed off in an unintelligible tangle of words. However, Bateman did catch the end of the statement: "... Freddie's Beach."

"It's an aircraft. That's Pan Am," Bateman stated urgently.

"I don't think so, sir," came the composed reply. "It's just a fire. It occurred on the way to ..." Again the speaker faded to indistinction.

"Well then, where is the Pan Am aircraft?"

"The Pan Am has still not come in as yet."

Oh, for Christ's sake! Bateman was now at the end of his patience.

"THAT WAS HIM," he screamed angrily into his microphone. "DO YOU UNDERSTAND? THAT WAS AN AIRCRAFT THAT CRASHED. WILL YOU GET TO THE SCENE IMMEDIATELY."

"Roger ... ah ... Pago." The previously calm voice at the other end was shaken by the blast. "But ... ah ... ah ... The two big crash trucks are now bound for that trouble."

Impact + 370 seconds. As soon as he had dressed, Tommy Bracken raced out of his front door and across the street to Saboro Fujii's house. He rapped loudly on the bedroom window and Fujii, half-asleep, opened it. Bracken told him to go straight away to the F.A.A.'s station at the airport. An incoming Pan Am machine was thought to have piled up and he wanted him to make a complete inspection of all the ground-to-air navigational equipment to see whether it was working.

Bracken then bounded back across the road to his car and drove off in the direction of the airfield. On the way, he decided to stop at the chief engineer's home and tell Pedro Pascua to check the flashing hazard beacon on top of Logotala Hill.

Meanwhile, the group of six survivors—Smith, Culbertson, the two Canns and the Merrills—sat huddled together in the damp, dark, tiny clearing, with the heavy rain dripping down on them from the dense foliage above, bringing a heavenly relief to the severe burns from which they all were suffering. They mostly talked about where they might be in relation to the airport. They assumed that the runway must be somewhere close by. But since the night was moonless and the bush relatively impenetrable, it occurred to some of them that they may have to stay huddled together right where they were at least until dawn, if not well into the next day.

Every so often the fires around the remains of the 707 would flare up once more. Dick Smith and Roger Cann decided that the plane still might explode. Smith told the others that he knew of a better spot a little further away from the wreckage. So again he picked Mrs. Merrill up in his arms and blindly led the group deeper into the jungle.

Meanwhile, the six duty firemen in the airport's brigade headquarters were jarred awake by the din from the klaxon alarm. They rapidly put on their protective clothing and reported to the downstairs duty officer's room. Sainila Fanene informed them that the R-1 engine, standing by for the arrival of Pan Am's Clipper 806, had seen a blaze past the end of Runway Five. One of the new arrivals, thirty-year-old Manu Moefili, asked what type of a fire it was. Fanene replied that they were not too sure. More than likely it was a house.

Impact + 500 seconds. It took Smith only a short time to stumble on another small clearing in the jungle—this one about twice as far from the burning Clipper. The area was circular in shape with a banana palm in the center, ideal as a backrest. He again put Mrs. Merrill down, leaning her against the tree. As the others settled, Smith and Roger Cann, quietly conferring, agreed that surely the airport people would have realized their plane was missing. Surely help should be on its way, if it was not already close at hand. They decided that one of them should return to the burning wreck to contact the fire and rescue personnel when they arrived, and the other should try to find assistance elsewhere.

Smith started off in a direction roughly parallel to the line of the crash path. Roger Cann made his way back to the aircraft along the narrow jungle trail they had forged through the undergrowth, gently, painfully brushing the vines and bushes to one side or the other, past the old resting place. He planned to approach the edge of the fire once more and then move around it to near the plane's nose. Suddenly he stopped, his attention caught by the unmistakable sounds of engines being revved up somewhere in the distance. He could hear them clearly above the surrounding noises: the constant patter of the dripping rain; the hissing, sizzling splutter of the wild inferno. After a few moments, he pushed on faster toward the still-burning wreck, assuming that any rescuers would be heading directly for it.

He completed almost a full circle around the edge of the holocaust, watching and listening intently. Now he could hear nothing other than the roar of the flames intermixed with, every so often, a mild explosion. The eruptions seemed mainly confined to the area next to the left wing; the right one was completely engulfed in steady sheets of fire. The passenger cabin was still intact, although the central part of it now was being melted away. The nose and tail were relatively untouched; they were slightly blackened and small clumps of flames licked around both. As far as he could make out there appeared to be no material damage to the airframe other than that caused by burning. He could see no gaping holes, no twisted metal such as might have resulted from a crash. He even noticed that none of the doors was opened.

Impact + 600 seconds. As soon as the engines had been warmed up, the drivers of the two big tenders, R-3 and R-4, prepared to leave their base. R-3 was a Walter. It carried 150 gallons of foam and 1,500 gallons of water. R-4 was an O-11-B, with 200 and 2,500 gallons of foam and water, respectively.

The R-1 truck, still awaiting the arrival of the Pan Am Clipper out by the runway, was an Ansul light transporter, holding 120 gallons of foam and water already premixed, with another 450 pounds of dry chemicals.

Fiaoo Masina decided to wait no longer for the 707. He started the engine and, after a brief warmup, headed in the direction of the fire, driving down the taxiway that paralleled the runway, just in case the aircraft did happen to land.

Back in the terminal building, Robert Sievers, Pan Am's representative in Samoa, again looked at his watch. He was becoming concerned. The flight was now a few minutes overdue. He walked toward the air traffic control office to learn the latest information.

Just as the trucks were about to leave fire headquarters, James Prendergast and Jay McLean also decided to have a talk with Frank Bateman.

Cary Wade finished his coffee in the airport's dinette. Now he knew something serious was afoot. He decided that Pan American's Operations Room would have up-to-the-minute news.

Impact + 620 seconds. Roger Cann continued his walk around the burning Clipper. Seeing no rescue personnel on the far side, he returned to the left of the airplane in case they might be arriving in that direction.

Impact + 700 seconds. Fiaoo Masina and Laalaai Niko drove the R-1 rapidly to the end of the taxiway and then off it, onto a small track carved by the engineers' service vehicles through the low-lying bushes. This was an access route out to the "middle marker" and beyond, up the side of Logotala Hill to the hazard warning beacon and navigation transmitter. They had not gone far along it before they came to a chain which was stretched across the lane and, at either end, fastened to short posts. The barrier had been installed by the Government of American Samoa as an excuse for the perimeter fence the F.A.A.'s regulations required and that they had promised to erect around the airport. It was positioned here because, a few hundred feet further on, the service track crossed a public thoroughfare—the main road leading to Freddie's Beach. The chain was of fairly solid steel, partly rusting, and at one end there was a large padlock holding it firmly to a cast-iron hook embedded in the post.

The men stopped, got down from the truck, and examined the lock. Niko returned to the vehicle and grabbed a heavy fire hose nozzle. He ran back, raised the thing in the air and slammed it down on the chain, trying to smash it. Twice, three times, four times, he brought the weighty brass down on the stubborn links. It was no use. The bouncing chain remained, unyielding.

Meanwhile, R-4 was leading R-3 in the rush along the same taxiway the other fire truck had taken moments before.

Impact + 800 seconds. The R-4 turned off the end of the pavement onto the service track and, reducing its speed slightly, continued until, rounding a bend, it came upon the bright red stoplights of the parked R-1. Laalaai Niko trotted back, explaining excitedly that they did not have the key to open the padlock holding the chain. Manu Moefili, in charge of R-4, turned to his handlinesman, Michael Tavai, and told him to get out and unlatch the obstruction.

Tavai ran to the barrier. After a quick inspection, he returned and took a heavy hatchet from the R-4's utility compartment. He raced back and began hacking ferociously at the hook-eye onto which the lock was attached. After a dozen or so solid swipes, he smashed it. The chain fell limply to the rain-soaked earth and the three trucks—R-3 had by now caught up—moved off in file. They soon arrived at the throughway. With warning lights flashing, they sped across the road. Moments later, lead vehicle R-1 again ground to a halt. The other two pulled up behind.

Masina had been stopped by yet another locked chain spread between posts on either side of the lane. Tavai was again called upon to clear the hindrance. While he was doing so, Masina conferred with Moefili and they agreed that the R-4 should take the lead the rest of the way because, as they went deeper into the jungle, the road narrowed and his truck had the greater fire-fighting capacity.

With the second barrier down, the R-1 moved forward a short distance before pulling over to one side and allowing the R-4 to pass.

Impact + 900 seconds. Roger Cann, now standing somewhere near the tail of the burning aircraft, suddenly noticed lights in the distance. They seemed to be approaching from two directions—from the area where Dick Smith had gone to search, and from directly in front of him. He again walked around the side of the wreck, retracing his path toward its nose. He was halfway along the fuselage when the R-4 came to a halt about sixty feet away.

The lead fire truck parked on the right side of the narrow gnawed-out lane on the very edge of which the front of the airplane rested. Manu Moefili, in charge of the vehicle, instructed Michael Tavai to unroll the main hose and direct the foam and water mixture onto the forward left-hand door area of the Clipper, which was still, even now, fairly intact. The fiercest fires seemed to be concentrated in the middle section over the wings.

Tavai crawled down the slope into the tiny ravine, dragging his line behind him. As he reached the bottom, the foamy water started to spurt from its nozzle—someone back at the truck had unexpectedly turned the cocks on. He directed the white soapy stream onto the side of the fuselage between the wing and the closed front passenger door. The flames continued to leap up and occasionally great balls of fire would shoot high into the dark, rain-drenched sky.

Pepe Mitaina, the other handlinesman on R-4, also climbed down into the gully, taking up a position slightly to Tavai's left, nearer the nose.

R-1 had pulled up a short distance behind the R-4. Its two crewmen prepared and then turned on their high-pressure hose. But before much fire-fighting could be done, the pipe burst, rendering the operation useless. They closed the cocks, put everything away, and walked down the incline and over to their fellow firemen from R-4.

Tavai could just hear human noises—weak, agonized cries for help intermixed with the crackles from the flames. They seemed to be coming from his left, somewhere close by, possibly on the side of the embankment near the burning cockpit.

He handed his hose to Niko, then gingerly trudged through the low-level brush in the direction from which the calls were originating. He climbed slowly up the steep, slippery slope and was only a couple of feet from the edge of the service track when he discovered the co-pilot lying in a pool of diluted blood, half-buried in the scorched undergrowth. Tavai bent down and asked him how he was feeling. Phillips said his left leg and right arm were hurting very much. Tavai tried gently to lift him. He decided against it. The immediate terrain was treacherous and Phillips was in intense pain; he didn't want to risk dropping him. He called over to Mitaina to come and assist. The two of them then lifted Phillips out of the blackened remains of the bushes and carried him to the deserted cab of the R-1.

Tavai returned to his position and told the R-1 crewmen that Phillips had been placed in their vehicle. They quickly went to the truck and backed it down the service road on their way to the hospital.

Roger Cann, in the meantime, had struggled up the side of the embankment to the edge of the track. As he scrambled up the last part, he was nearly run over by the then reversing R-1.

Moments later, several policemen came trotting up the lane. Cann called to them. Gesturing, he tried to tell them there was a party of survivors in the bushes. The lawmen stared incredulously at his dishevelled body—much of Cann's clothing had been burned away. They just stood there, gaping, talking among themselves in Samoan. Cann, pleading with them to come, started to hobble away. A few seconds later, the police decided to follow.

Soon after R-1 had departed, the R-3 drove up the service track and stopped just behind R-4.

Impact + 1,100 seconds. Meanwhile, Smith had reached Freddie's Road. It had been a long, hard grind through the dense undergrowth, and even though it was only about 100 yards away from their mid-forest clearing, he had at last found it. He paused to rest by the edge of the thoroughfare.

Several minutes later, a jalopy came hurtling along it. Screeching to a halt nearby, three Samoan boys jumped out and one came racing back to Smith, saying "Can we help? Where are all the people?"

"Well," Smith replied slowly, exhaustedly, the water dripping down his face and intermixing with the blood still seeping from the cut on his forehead, "all the people are down there in the jungle. They need lots of help."

The trio immediately started off into the bush in the direction Smith had pointed.

"Hang on a minute," Smith called to them. He knew they would have trouble getting to the group, and he noticed that the driver was only wearing a pair of short shorts and no shoes.

But his plea was in vain, so he followed a brief distance before deciding to look for further assistance. He turned, went back to the road, and began walking along it.

At the airport's fire station, a crowd had gathered. Among them was Robert Sievers, Pan Am's local manager. As soon as the R-4 had radioed confirmation that it was an aircraft that was ablaze at the end of the runway and not a house, Sievers and Max Schwanke leaped into a Pan Am van and drove off toward the scene. They were followed in file by Joe Lion and Vinne Atafu—two other of the airline's employees—and by a police car that had pushed its way into the convoy. Next in line was Cary Wade's truck and, at the rear, South Pacific Island Airway's manager, George Wray, in his car.

As he walked along the side of Freddie's Road, Smith

rounded a small bend and saw some vehicles coming toward him. Their headlights were glistening off the wet pavement. A couple of them had flashing beacons on their roofs.

Several of the cars shot past Smith, evidently not noticing him. The last in the group was a gray or white van—in the darkness and pouring rain it was impossible to tell which. It ground to a halt. Smith scurried over to it. A man jumped out and Smith asked him whether he was from Pan Am. Yes he was, Schwanke replied, and inquired where all the passengers were. Smith said he didn't know where *all* the people were, but he could certainly lead them to some. Would he happen to have a flashlight? Schwanke did. So Smith started to direct Schwanke and Robert Sievers down to where he had left the little group of survivors in among the dripping vegetation.

Moments later, Heather Cann and the others had been found by the three Samoan boys. The lads were standing, staring, and talking away to one another in their native tongue. Heather pleaded with them not to touch her. She was badly burned, she said, and it was very painful. Soon her husband and the policemen joined them.

Roger helped his wife to her feet, supported her from behind, and the two of them followed one of the boys back up the path through the jungle toward the highway.

Impact + 1,150 seconds. When the remainder of the auto convoy arrived at the intersection of Freddie's Road and the service track at just about midnight, the place was in utter confusion. The wind had picked up again and was gusting with great force. The rain was pelting down. Cars, trucks and vans were parked indiscriminately all over the area. The remains of the aircraft was ablaze from one end to the other, with the most furious of flames still concentrated in the middle. Minor explosions were occurring spasmodically—presumably "pockets" of fuel blowing up.

Back at the jungle clearing, Max Schwanke told Smith that he had better return to Freddie's Road and get medical attention for his burns. This was the first time Smith had thought about this subject. In the dim, undulating light he could just make out that something was wrong with the skin on his hands. As he looked, he could see blood dripping on to them from somewhere on his head.

One of the other people standing around suggested Smith come with him; he would take him to the hospital. He was a local guide who had been waiting at the airport for Clipper 806 in order to meet a party from Sun City, Arizona. Sun City was not too far from Smith's home. As a matter of fact he knew it well. The driver said he was to collect a group of six (possibly those in the row behind the Canns during the flight) to take them to their hotel for the night and for a tour of the island the following morning.

Smith walked slowly behind the guide, up to the main thoroughfare. When they arrived at his vehicle they were amazed to find Roger and Heather Cann sitting in the back seat. The Canns had forced their way to the road and had entered the first car they saw in order to get out of the soaking rain. Smith and the guide got into the front of the automobile. The next problem was, they were blocked in. While the driver got out and cleared a path, Smith also climbed out and went to another vehicle that had its headlights on in order to take a closer look at his wounds. His hands were very puffy and white. His clothes were not too badly burned. While he was checking himself over, a young Samoan girl approached and, opening her handbag, offered him half of a broken rear-vision mirror. Smith smiled as he accepted it. He observed the reflection of his face. The cut above his right eye was bleeding profusely. *"Good,"* he thought, *"that means there should be no infection."*

Then the guide called him back. The way was now cleared. Smith handed the mirror to the young girl and returned to the vehicle. No sooner had he closed the door than they took off.

They were racing at high speed along the muddied, nearly deserted country road. The torrential rain poured down.

Smith was being scared out of his wits by the young man's driving. A couple of times he leaned over and begged him to slow down; he didn't. So Smith finally reached out and whipped the keys from the ignition. The action practically scared the driver to death. As the car was rolling to a halt, Smith explained with some agitation they they had already been lucky enough to escape from one accident—they didn't want to be involved in another. He handed the keys back. But it had done no good. After starting the motor, the man again plunged his foot to the floorboard.

While the fire-fighting efforts were continuing at the crash site, back at the F.A.A.'s office at the airport James Prendergast was talking with the local manager, Tommy Bracken. As soon as they had reported the now-confirmed accident to the F.A.A.'s Communications Center in Washington, D.C., they left for the scene to see if they could help.

They arrived at the intersection of the two roads just as the police were gently loading the other located survivors into their automobiles for the eight-mile dash to the hospital.

7

After parking their car near the scene of the crash, the senior employees of the Federal Aviation Administration, Tommy Bracken and James Prendergast, strode briskly toward the burning Boeing 707. They circled its nose for a closer look at its right side where clouds of thick smoke were billowing from a slight break in the mid-section of the fuselage. This was accompanied every so often by what appeared to be spouts of steam. The flames still occasionally shot skyward from various places, and there were still spasmodic explosions.

Having inspected the heat-caused cracks, they decided to walk to the left side of the plane where most of the activity was taking place. They circled the nose once more, skirting it along the service track on which the very front section was lying. On the other side, the flames had sufficiently died down for them to approach the cockpit windows. They could see nothing inside. They then moved a little way into the gully—only a few paces—and Bracken tried pulling open the forward passenger entrance by its handle. The metal was extremely hot. He could move it only slightly. And even though the surrounding area had been blackened by the fire, a careful inspection revealed that the door itself was unlocked—it was not quite flush with the fuselage. He tried pushing it. Something inside, however, seemed to be leaning against it. It would move very slightly inward and then spring back to its original position. Wanting to see what was stopping it, he looked through the cabin windows just to its rear. The interior was pitch black. He could not see a thing.

He then walked down along the left side of the Clipper further into the gully, joining the police and firemen who were searching for the burned bodies of any passengers that might be there. Scouting quickly through the scorched remains of the jungle in that area, they found nothing, so Bracken returned to his car and drove back to the airport to give the F.A.A.'s headquarters the latest information.

Prendergast remained at the scene. As he continued his tour of inspection, he concluded that the whole interior of the cabin was ablaze. The fires, hidden from direct view by the dense, acrid smoke, were quite evidently spreading along the rear portion and up the side walls. He knew that some survivors had been located in the woods, but decided that finding any more would be almost impossible without the use of both floodlights and megaphones.

One of the other searchers in the little ravine near the left of the airframe was Cary Wade, the air-taxi company's mechanic. He looked up at the wreckage. He, too, noticed that the flames inside the 707 had evidently increased in intensity and were now burning the whole of the roof area away. The heat forced him and the others close by him, to retreat into the jungle's undergrowth, where he continued his hunt. He found no one.

At a quarter past midnight, Pan Am's Sievers arrived back at their operations room. There, he asked another airline employee, Linda Sweitzer, to telephone his wife and have her bring over some blankets for the survivors. He reported the accident to the company's head office in New York and then returned to the crash site.

In the meantime, George Wray, the owner of South Pacific Island Airways, had also arrived at the scene. By then the rainstorm had died away to a drizzle, taking the wind with it. Flames were shooting straight up into the air to a height, Wray estimated, of about fifty to sixty feet. He walked over to the nose, trying to see into the cockpit. As he approached, he noticed a large hole in the right side of the pilot's area. He bent down and looked through it. He could see nothing in there. No flames. Nothing.

He then plodded through the mud around to the left-hand side of the plane where the group of people in the gully were still searching for survivors. Striding down to them he encountered a young Samoan boy who, pointing excitedly into the charred ashes at something just beyond the overhanging side of the fuselage and behind the port wing, told Wray he could see a body. The heat was intense, but just bearable. The fire had been extinguished in the area where the corpse lay, but there still were a few small flames coming from what appeared to be white-hot glowing metal. The rear part of the wing had been burned away and in the immediate vicinity there remained, here and there, some smoldering clumps.

Both Wray and the lad could see the body, but they couldn't get near it because of the temperature. It appeared to be the remains of a young Samoan girl, about twenty years of age, lying face upwards beneath the outer curve of the rear cabin, her feet pointing toward the wing.

At around half past midnight, George Wray decided to comb the area of the crash path. There was the possibility that some of the passengers might have fallen, jumped, or been thrown from the skidding Clipper. Other people had the same idea. It took the group about three quarters of an hour to complete the search. Again, no one was found.

At the airport, James Prendergast tried to get an urgent message through to his home base, Honolulu, to let his office of the F.A.A. know about the accident. The operator refused to put him through immediately, insisting that she would call him back when it was his turn. He tried time and again to change her mind. Finally he simply placed the call, awaiting her pleasure for the connection.

Meanwhile, he gathered together all the flashlights and Hawaiian flares he could find at the airport and returned with them to the disaster site. Leaving them for the rescuers who were searching along the crash path, he went back to the office in the terminal to await his telephone connection with Hawaii.

Suddenly, startlingly, the wreckage let out a loud "crack." A shower of flames and sparks shot high into the dark midnight sky. The back of the Boeing 707, melted and weakened by the tremendous heat, unexpectedly gave way and collapsed into the gully below.

Smith and the others, meanwhile, had been taken to the fracture ward of the L.B.J. Hospital for Tropical Diseases in Pago Pago. They were in what was, normally, a visiting room for Samoan parents whose children had suffered broken bones. The room had been turned into an initial "inspection and interrogation center" for the burn victims from the air disaster. The full medical staff had been alerted by the earlier telephone call from June Bracken, and were now ready for what they presumed would be a huge influx of injured.

As they had entered the ward, Smith and the two Canns had immediately been met by doctors and nurses. The man in charge, a short, wiry, very efficient medico named Ecklund, asked Smith how many might have escaped from the wreckage. Smith said he didn't know the figures, adding that he felt sure almost everyone must have survived. He told the doctor he was of the opinion that he should prepare immediately for a large group of people; on being pressed, he suggested that there could be around one hundred.

Smith was told to sit down on one of the many chairs the hospital's emergency staff had placed in a long corridor. Where the passageway widened, a few make-shift beds had been quickly arranged, separated from direct view by some rapidly strung-up sheets.

As he waited, Smith began getting fidgety and started moving about. As a walking, talking patient, he was in less immediate need of urgent attention than some of the others. At that time, the most experienced medical men had the terribly burned co-pilot, Jim Phillips, in an emergency room, attempting, to the best of their ability, to clean up and evaluate his wounds. The next most serious case seemed to be Heather Cann.

Finally a senior nurse, accompanied by a group of assistants, came over to Smith and, taking him into a small room, washed and cleaned his injuries and then gently led him to a long, narrow twin ward. He was helped into the bed near the window. As at the airport, there was no glass in the opening but only a thin screen behind a canvas curtain. Everything about the infirmary was sanitary and efficient.

Smith was suffering from third-degree burns to his hands and face and both in and on top of his right ear. The deep gash above his right eye was also badly singed. At some time during his getaway he must have strained his neck. He couldn't remember when, but it had given him a great deal of trouble right from the start. His back also was causing him difficulty. So was his right leg. There was an extremely sore spot on his right knee which was very painful when touched.

Within seconds he had fallen asleep, either from the shock or from the effects of the pain-killing and other drugs he had been given.

By two o'clock in the morning, the blaze at the wreckage site was mostly under control. Bob Sievers of Pan Am approached the tail to see how the fire-fighting had progressed in that area and noticed, at the very rear of the passenger cabin, that the flames had been completely extinguished. He walked over to the left side exit door and, grabbing the handle, made several attempts to open it. The thing would not budge. It was still quite hot and he could not keep his fingers on it for more than a second or two. Looking around, he saw that the main fuselage walls had been burned away almost to the level of the windows. So he climbed over the top and lowered himself gently inside. He was just forward of the rear-left exit in what had been the auxiliary galley. His idea was to retrieve the flight data recorder—the "black box"—positioned in the very aft section under the tail. As he slowly let himself down on what had once been the galley floor, his feet landed on something soft. Peering, he could clearly make out two blackened bodies. Both were lying with their backs uppermost, their heads toward the aircraft's nose. In the flickering of the remaining flames from the still-afire mid-cabin section, Sievers could see that the whole of the interior metalwork at the very rear of the airplane had been melted and distorted. This, he realized, would prevent his reaching the position in which the recorder was housed. He decided to make another attempt at recovering it later from the outside.

As the flames died away in the nose area of Clipper 806, rescuers, using a flashlight, peered through the cockpit from outside the windshield. As the beam slowly panned around the blackened, partly liquefied remains of the control and equipment panels, it revealed the silhouette of a man—the charred body of Captain Pete Petersen—still sitting erect on the frame of the pilot's seat. There was no evidence of Petersen's having moved since he had undone Phillips' safety belts, then turned and watched the other members of his crew disappear into the darkness of the cabin. For some strange reason he had made no attempt to follow Flight Engineer Green or First Officer Gaines. Neither had he tried to get out of the hole in the cockpit wall through which Phillips had fallen.

Smith was aroused from his deep sleep by some activity

in and around the other bed in his hospital ward. He couldn't see what was happening because someone had drawn a curtain between them. He listened. One of the doctors was saying, "Mr. Phillips. We know you can't talk to us, but could you nod your head or signal your permission for us to amputate one of your legs?"

There was a pause.

The speaker continued, "You will not live unless we operate. And we can't promise you will even if we do."

The two medical men gently, calmly, continued to press the co-pilot for his sanction.

Soon, Smith again lost consciousness. As he slid away, his mind returned to about 1930, when he was a freshman in high school. He was just learning to dive. He was with a group of friends at Riverside Park in Phoenix, practicing plunging from a platform 48 feet above this eight-foot deep swimming pool which had been built on the river's bank and was lined with boulders.

He had made the dive successfully on many occasions. This time he was instructed to roll his body over just after he hit the water—it would stop him from going too deep. Smith tried it. He did not quite succeed. His head collided with the base rocks. Someone saw he was in trouble. He was dragged from the pool with a fractured neck and a severe concussion.

Now in the Samoan hospital as he sank into oblivion it all came back to him: those vivid reds, greens, and blues that swam in his mind as the collision took place; the grinding, snapping noise, then the pure and utter serenity, the peacefulness, the tranquility as all consciousness disappeared.

Fred J. Uhrle, the director of the Pago Pago Port Administration, had been awakened by a telephone call from John F. Zumdieck, the authority's special adviser. He was told that a Pan Am Clipper had crashed and was burning profusely near the airfield. Uhrle quickly dressed and left his house, arriving at the disaster site just as Petersen's corpse was being removed from the remains of the cockpit. Uhrle walked over to a group of other officials—Ronald Fano, the airport's manager; Li'a Tufele, the Commissioner for Public Safety; Arthur Ripley, the fire chief; Sergeant Silila Fateia, from the local police; and John Zumdieck. They discussed the fire-fighting and rescue operations' progress, and suggested that the Coast Guard commander have his men conduct a further search in the surrounding bushes and undergrowth for bodies or survivors. Very few of the people who had been on the flight had been accounted for; the others must be somewhere.

The management team then made arrangements for electric generators to be immediately—urgently—brought to the scene from the airport. They sent one of the policemen back to the terminal to call the Public Works Department's after-hours number for a further supply of the portable units.

The group next decided to assist the Coast Guard men in their search. They concentrated on the area close to the burned-out fuselage. They found no bodies, but did discover a number of cameras, binoculars, and distorted, but still easily recognizable, key chains. These were handed to Bob Sievers of Pan Am. They would be vital evidence in the positive identification of those persons who had been on board.

At about half past three, Sievers was joined by Prendergast of the F.A.A. and Wade, the air-taxi company's mechanic, who told them that they could at last approach the tail section of the nearly burned-out Clipper. Sievers went to his van and returned with a hatchet. He began to chop at the cone under the aircraft's tail, finally opening up the area inside which the bright yellow "black box" rested.

While Sievers had been hacking from the outside, Wade had again entered the rear passenger compartment, stepping over what remained of the side wall. The whole plane had been practically burned away by this time, almost to ground level in a number of places. He called that he could see a line of bodies in what had once been the central cabin aisle. Most were lying at an angle, with their feet to one side or the other and their heads toward the center of what used to be the passageway. Wade noticed two others at the very back, past the rear doors, in the vicinity of where the lavatories had been.

Sievers reached inside the hole he had cut in the section under the tail. After a struggle, he extracted the recorder and handed it to Prendergast. Fires were still slowly burning in the mid-fuselage area; every so often they would spurt up, then die down again.

The three men started back to the airport to lock the recorder in a safe. As they walked past the remains of the once-proud flying machine, they commented to one another that some of those aboard must have been overcome while they waited in line for an escape out the rear exit.

By then, the full horror of the disaster was just beginning to make itself known.

It was no use searching for survivors any more. There simply were none.

When the rescuers peered into the darkened, calcinated cockpit as the smoke inside cleared, they saw—just to the left but a short distance behind where the seated captain had been found—layers upon layers of corpses. As they stood waiting they must have simply collapsed to the floor, neatly stacking themselves one on top of the other like toppled dominoes.

By about four o'clock, the temperatures in the area had cooled enough for the rescuers to start the grim task of removing the victims' remains.

The territorial police, together with officals from the Pago Pago Port Administration, assigned one of the airport's disused buildings as a temporary morgue, somewhere to take the passengers' bodies for safekeeping. It was a small warehouse adjacent to the old runway. Two of the constables were detailed to keep a strict guard on the shack.

As the corpses were retrieved it was apparent they had been human because of their shape, but that was about the only similarity. They had been burned almost beyond recognition. They were carried by truck or van back to the make-shift mortuary.

Fred Uhrle of the Port Administration realized the forthcoming problems. He instructed the rescuers to be careful not to detach any brooches, watches, rings or other ornaments which could, later, assist with identification.

In the spreading cool of the early hours of the morning, they dragged nearly forty bodies solely from the area between the cockpit door and a few rows forward of where Dick Smith had been sitting, near the front of the wings. A like number was found in a line leading to the rear main entrance.

The mopping-up operations continued until daybreak. The morning's sun revealed a sight of enormous horror.

As the last of the human remains was being taken away, the police advised that they were two short—they could account for only 99 people; Pan Am had said there were 101 on board. A further search of the cooling wreck was called for.

John Zumdieck and Fred Uhrle decided to check the aircraft's crash-path for themselves. They walked along it, among the mowed-down greenery, as far back as the Eucalyptus tree. The whole area was scattered with debris and the latter half of the scythed-down strip was littered with pieces of clothing, books, papers, shirts, and other personal belongings which had come from the passengers' luggage, together with fruit, vegetables and cakes of all sort, released when the aircraft's underbelly was ripped open during the slide.

After returning to the crash site, Zumdieck and Uhrle instructed the airport's security guards to maintain a close watch along the strip, preventing people from moving or taking anything away.

Fred Uhrle excused himself from the rest of the management group. His watch showed the time as now half past six. He had duties to perform. He had to return to his bungalow to pick up his children and take them to school.

It was at this point that a head count of the rescuers was called for. All those who could be spared were sent home to try to get some sleep. A small number from each of the different teams were required to stay, either at the wreckage, the skid path, or the temporary morgue, maintaining security. The new fire shift arrived and was given the task of making sure that the flames which had been fought so long and so hard did not flash to life again.

At half past seven, a management meeting was held in the Pan American offices back at the airport. Present were Frank Mockler, the Acting Governor of American Samoa; Bob Sievers of Pan Am; John Zumdieck, the special adviser to the Port Administration; Fred Uhrle, the Administration's director; Arthur Ripley, the fire chief, and Tommy Bracken of the F.A.A., among others. With the heat of the new day's sun beginning to rise, their first concern was for the victims' bodies. John Zumdieck was assigned to call the local manager of the Pacific Far East Shipping Line and request the use of two of their large refrigerated containers. Permission was granted and airport trucks were sent to collect them from the main downtown docks. By nine o'clock, both had arrived.

Meanwhile, members of the overnight fire-fighting and rescue parties returned to their homes. After the enormity of the previous nine hours' activities, sleep was impossible for some. So they simply showered, shaved, changed their smoke-sodden clothes, and retraced their steps to the scene to see whether they could help any further.

As the tally was still two bodies short, a number of the returnees was given the task of conducting another search. (By mid-afternoon, one had been located. It was then decided that a miscount must have occurred—but the figures later confirmed that it was still missing. A further minute scrutiny of the fire-charred remains located the elusive corpse hidden beneath the fuselage in the bottom of the ravine on Thursday, February 7—more than a week later.)

Seven men from the airport's management division were assigned the grim task of transferring the bodies from the warehouse to the refrigerated containers. But before that could begin, the internal temperature of the large boxes had to be lowered to thirty-four degrees. By ten o'clock, they were ready. Following the instructions from the Public Health Department, the seven men started removing the corpses individually, wrapping them in three sheets of paper and tying each sad bundle up with masking tape. By three o'clock, this horrendous task had been completed.

Smith was awakened fairly early by the medical staff at the hospital. Following the normal start-of-day routine, he looked for his wristwatch. It was missing. He checked with the nurses to see if they had removed it. No, they replied, they hadn't. He must have lost it somewhere out in the jungle or during the escape from the burning wreck. Or, possibly, it could be in one of his coat pockets. He asked them to bring his sports jacket to him. There was nothing in it at all.

Smith, now, was not sure what was going on. He was positive in his own mind that he had placed some personal things in the inside breast pocket of the coat. If he was wrong, then he must be going mad. He lay back and started retracing the whole nightmarish evening in his mind from the take-off in Auckland to the present.

His memory arrived at that point where he was reinspecting the swimming pool plans for the Woodlands complex. He then recalled that he had not advised Houston of the error he had discovered, so he asked one of the hospital staff if they would send a cable saying that the construction contract should not be signed because of this mistake, but that he had corrected a copy of the drawings and he hoped it would be rescued from the ashes.

In the meantime, Pan American World Airways released the names of the passengers who had been on board to the press. They also attempted to list the persons who had survived. Both lists turned out to be incorrect and a totally unnecessary and frustrating turmoil ensued as a result.

At midday, the Federal Aviation Administration's technical staff had assessed that, other than for the squashed "middle marker" lying under the wreck, all of the airport's equipment was working "properly," and as the accident had happened away from the immediate runway area the airfield could again be opened for traffic.

The next time Smith awoke he realized there had been a change in the occupant of the other bed in his ward. The co-pilot, Jim Phillips, had been transferred elsewhere, presumably to an intensive care unit, and Roger Cann had taken his place.

After inquiring of one another how they were feeling and progressing, they spent practically the whole of that day reliving, second by second, their experiences during the preceding evening. However, after Pan Am released the passenger list, that afternoon they were both being constantly harassed by people, complete strangers, calling them on the telephone, or attempting to speak with them in order to ask questions.

The first thing the following morning a doctor, who said he had flown to Pago Pago from Honolulu the previous evening on the first flight in after the crash, pushed his way into their room and started talking with them. He said that his wife had been the senior purser on the destroyed aircraft, Gorda Rupp, and he was most anxious to find out what had happened to her. "Did she suffer? Did she die quickly?" he asked, "Or did she burn?"

Smith told him of his standing in the central passenger aisle and being knocked to the floor by the gases or other substance that swirled around his head. He suggested that, because Ms. Rupp was up front in the cabin, and because that was where the toxic material seemed to be originating, she must have died quickly. This appeared to satisfy the husband and he departed, evidently relieved.

At half past ten that morning—Friday, February 1—the

accident investigation team from the National Transportation Safety Board arrived. About an hour and a quarter later, a Pan American World Airways Boeing 707 freighter also landed. It carried a cargo of caskets in which the bodies would be transported to Los Angeles for tests and identification. Since the company had released the mixed-up passenger list the morning after the disaster, the airport's management staff had been visited by a number of local residents who had lost relatives in the holocaust. It was decided, because of the condition of the charred remains, not to allow anyone to view them.

In order to save further anguish, it was also agreed that the corpses would not be touched again during daylight hours, just in case the activities might be seen by the distraught families or others. They planned to unload the caskets from the freighter starting at ten-thirty that evening.

By half past midnight the following morning, the individually filled coffins were aboard the Pan Am 707. As soon as tie-down checks had been completed, the aircraft's engines were started and the machine slowly taxied out to the end of the runway. At just before one-thirty, the Boeing lifted off the asphalt surface, flying, as it climbed, directly over the tangled, burned-out mess that had once been called *Radiant*.

The next morning, a Samoan policeman entered Smith's hospital ward. "Mr. Smith," he said, "these papers had been found out in the jungle." He handed over a small packet. The contents were still damp, soaked by the torrential downpour of two nights previously. Smith opened it. Inside was his passport, his wallet, and the letters of introduction to the people whom he had been intending to visit in Western Samoa. The incident boosted his morale no end.

The week after the accident, Smith was fed up. He was fed up with being confined within a limited area, and he was fed up with the strictness of a hospital routine. He begged Dr. Ecklund to be allowed to stay at the only big hotel in Pago Pago, The Malaeimi, promising that he would return to the outpatients' ward each day. Following a conference with the other physicians, Dr. Ecklund told Smith that the next morning Pan Am's so-called Corporate Medical Director, a Dr. Joseph G. Constantino, would give him an examination and if he seemed to be all right he would be released.

The following day, Smith was out of bed early, washed, dressed, packed, breakfasted, and eagerly awaiting the arrival of Constantino. The man did not turn up.

At four o'clock in the afternoon Smith decided that if he was not there by five, he was going to leave, anyway. At five minutes to five, with still no sign of Dr. Constantino, Smith telephoned Pan American and told them that he had waited around all day for their medical specialist and he had not appeared, therefore he was checking himself out and could be located at the hotel. The Pan Am employee told him he could not do that. Too late, Smith replied, he had already ordered a taxi and it was due at any moment.

As Smith walked into the lobby of The Malaeimi Hotel, the first person he saw was Constantino. He had a half-filled glass in his hand. The doctor immediately recognized Smith and approached him. He apologized for not coming to see him that morning as had been arranged, claiming that he'd been so busy he had been unable to find the time. Smith looked down at the cocktail the doctor was holding. Constantino added that he was glad he had checked himself out of the hospital, and he would give him the medical examination any time that suited. Smith suggested that they wait until the next morning.

One of the people with Constantino then asked Smith whether he would care for a drink. Smith replied that he would like a Scotch. A few minutes later a member of the group said they were planning to have an informal meeting in one of the hotel rooms upstairs and invited Smith along for "a chat."

Later that evening, after dinner and a rest, Smith ventured up to the room in response to the invitation. Soon after he had entered he realized that the "friendly social gathering" he had been led to believe would be taking place was not as "friendly" as had been made out. It was more of a solid, abusive grilling. There seemed to be more Pan American people there than anyone else. Every time Smith was asked a question and answered it, a Pan Am employee immediately would either start presenting excuses or he would belittle Smith's knowledge of the situation or his ideas.

Following one particularly heated attack from a Pan Am staff member, Smith stormed out of the room. He had had enough. He was completely fed up with both their attitudes and their insincerity. He suspected that either the F.A.A. or the N.T.S.B. had secretly tape-recorded the proceedings; he hoped the cassettes would be listened to by higher-ups in either organization and that they would take action when they heard the way he had been treated. (Both the F.A.A. and the N.T.S.B. later denied having recorded the conversations.)

Since he had released himself from the hospital, the next thing Smith wanted badly to do was to revisit the scene of the crash. A number of people had said they would drive him there so he could have a look at it. None of the offers was ever carried out; there was always some excuse. Finally he cornered an F.A.A. official and made him promise to take him.

The next morning they departed. They sped in one of the F.A.A.'s hired cars along the road toward Freddie's Beach. As they neared the intersection with the engineers' service track, the driver slowly came to a halt, pulling off to one side of the thoroughfare as far as he could.

Smith, still wearing the tattered but now clean remains of the clothes in which he had escaped, was led back up the access route, past the chopped-down chain, to the scene.

He later recalled what he saw:

"When we got to the plane, I was really shocked. The right side of the aircraft was nothing but a big gob of ashes looking somewhat like a giant ashtray. Parts of the other side of the fuselage still existed as I had seen it that night in the jungle.

"We walked back along the crash path. You could see from the ground burning that the fires had started at approximately the same time as the engines had dug into the earth.

"The investigators showed me the different pieces of undercarriages and wings and other parts of the machine that were all the way up and down that stretch of jungle. I could see the remnants of clothing, the smashed-up pieces of luggage and the remains of what appeared to be tons of vegetables scattered and drying out in the tropical heat like a huge, unwanted minestrone."

Smith lingered at the site for approximately an hour before deciding he had had enough. Then he wanted to get out—to get right away from the place—to get back to the States and forget about every aspect of the event.

On his next visit to the hospital, he pleaded with the doctors to be allowed to return home. Since his injuries were healing well, they decided that soon he could go.

But how was he to get back to Phoenix?

This was the middle of the South Pacific. His choice was by sea, or by . . .

"Oh, no!"

Smith doubted whether he could ever enter an airplane cabin again.

One day, as he slowly strolled about the picturesque town of Pago Pago, killing time until the medicos would give him a release to return to Arizona, he entered a tiny park near the waterfront. There was the rickety old cable car jerkily, slowly weaving above the bay and up the side of Mount Alava. He sat in the sun for a few hours, watching the ancient object wend its way backwards and forwards.

"If I could go on that thing," he thought to himself, *"I reckon I would be able to fly once again."*

He made arrangements with a Samoan boy who worked

as a tourist guide to accompany him, just in case. Plans were set for a taxi to pick up the two of them the following morning. With Smith and the guide in the back seat, it left the hotel and started off in the wrong direction. Smith asked why. The driver told him it was a new cab and he didn't want to take it up the steep Solo Hill from where the cable car started. He was just going back to his base where they would transfer into an older vehicle.

Once they were in the other taxi, Smith started to worry. The driver said that he would change over to an "older vehicle," not one *that* ancient.

By the time they had made it halfway up the steep side of Solo Hill, both Smith and the guide were ready to forget the whole thing.

When the antiquated taxi finally arrived at the site where the cable car started, there was a crowd of people standing around. They had had problems with the device that morning. Twice the mechanism had jammed, leaving the passenger cabin swinging precariously in mid-air until the fault could be corrected and the motors restarted. Now Smith had an excellent excuse to back out.

They admired the view for a few minutes. Many other people who had been standing around had now left. Smith told the boy he would just go into the office and ask how long it would be before the trouble was corrected. He decided that if the delay was going to be longer than, say, five minutes, he would forget it altogether.

Moments later he returned to the viewing platform with two tickets in his hand.

The few remaining tourists were hustled into the aerial cabin, the door was closed, and jerkily, tentatively they took off.

Somehow they made it all the way up to the terminal near the top of Mount Alava. But as the front of the cab entered the station, the mechanism suddenly ground to a halt. Just a few feet more and they would have had a normal arrival. The attendants had to force the car's door open and, using outstretched arms and ropes, physically haul the thing into position.

Smith and the guide walked the few hundred feet to the lookout point on the very summit of Mount Alava.

The view was breathtaking.

The slight difficulties experienced in the final few feet of the ascent were quickly forgotten.

Smith now realized that if he had mustered the courage to ride in that thing, he could even enter the fuselages of the big jets without fear.

Of the 101 persons who had been on board Clipper 806 as she gently bellied into the Polynesian jungle, only ten actually got out of the plane alive.

One passenger, found in the bushes at the very edge of the fire, died the next afternoon as a result of his burns.

On the Saturday morning following the crash, co-pilot Jim Phillips' condition suddenly worsened. The doctors decided it was essential that his badly burned and broken leg be amputated. The operation was performed. He died the same afternoon.

Mrs. J. Merrill, who everyone thought was making progress, collapsed and passed away that Saturday evening. So did two of the other rescued passengers.

Because Mr. Merrill had made such a good initial recovery in spite of his terrible injuries, the doctors decided he could be transferred to the San Jose Burn Center about fifty miles south of San Francisco, on the morning of Friday, February 8, 1974. Merrill never reached Honolulu alive. He expired on the plane during the flight.

That left just four persons—Smith, Culbertson, and the two Canns—who, after weeks of intensive care and months of recuperation, survive today.

8

It was at the point at which Clipper 806 ended its rapid slide through the jungle that bureaucracy started, and, inexplicably, soon afterward stopped. An accident had occurred and the worldwide agreements state that it had to be officially investigated.

The definition of an accident according to the International Civil Aviation Organization (I.C.A.O.), is "An occurrence associated with the operation of an aircraft which takes place between the time any person boards the aircraft with the intention of flight, until such a time as all persons have disembarked, in which:

    a. any person suffers death or serious injury as a result of being in or upon the aircraft or anything attached thereto, or

    b. the aircraft receives substantial damage."

Of course, that means that they are interested only in the plane itself; the inquiries are conducted into the reasons for the machine's ending up where it did. The only objective of the investigators is to say that because of their findings they trust a similar event will not recur in the future; yet that is exactly opposed to reality, most of the time.

Neither are they too interested in how or why death or injuries were inflicted on the passengers. Any questioning of the survivors is purely to assist them in determining the causes of the damage to the airplane.

And each aerial calamity, after such fragmentary research, is expected to result in the issuance of a documented summary:

"[Inquiry] The processes leading to the determination of the *causes* [italics added] of the accident including completion of the relevant report."

As can be seen, they are not interested in the *effects*.

In point of fact, Pan American World Airways, in January, 1974, was in the middle of two crises. One was monetary. And it was midway in a nine-months' period during which the company lost a quartet of its Boeing 707 fleet, all to disasters.

Of the four-in-a-row accidents, the first was on the evening of July 22, 1973. The plane had just taken off from Papeete, Tahiti, and during the latter part of, or immediately after making a wide, sweeping, left-hand turn out over the Pacific in order to avoid flying above the sleepy town, it was seen in the fast-fading daylight by a fisherman to come gently down and crash into the ocean, sending up a huge plume of sea-spray.

Seventy-nine persons were aboard. Seventy-eight of them died. The lone survivor, a young Canadian, had been seated in the very forward left-hand side of the economy section. He remembers his attention being drawn by something bright and flashing in the vicinity of the port wing. The next thing he recalls is being dragged semiconscious from the water by the first rescuers on the scene. He knows nothing more.

The aircraft sank to the ocean's floor. The French are

responsible for the investigation, and, in their usual fashion, at the time of this writing they have released no information at all. It is doubtful, however, that they have much of a story to tell, because they decided, for reasons best known to themselves, to give up the task of recovering the wreckage.

The second accident occurred on November 3, 1973, to Pan Am's Boeing 707 at Boston, as we have already described.

The third was the mishap to Clipper 806 at Pago Pago.

And the fourth happened at Tinga-Tinga on the island of Bali, Indonesia, at 2326h on the night of April 22, 1974.

The aircraft was on a flight from Hong Kong to Sydney, Australia, with a scheduled stop at Bali. At 2228h, the captain was cleared by Jakarta air traffic control to descend from cruising level to a height of 19,000 feet. At 2308h, the crew spoke by radio with the Bali controller and told him they should be landing at 2327h. They never arrived. It was later discovered that they had collided with a mountain approximately forty miles to the northwest of the airport.

The wreckage landed in among forty-five- to sixty-five-foot high trees. From the swath cut through them it was determined that the flight had been banked to the left. The right wing tip initially smashed its way through several large tree trunks, lopping them off at a height of fifty-five feet above the ground. The right wing then collided with a massive tree and was sheared off, breaking into four large sections as it did. The left wing struck a ridge and likewise was torn off. The remainder—mainly the fuselage and tail—plummeted into a cliff face. After impact, raging fires erupted in the different disintegrated pieces.

The cause of the crash was reportedly uncovered on the rescued cockpit voice recorder, a dictaphonic device containing a tape-loop, which picks up the general cockpit sounds together with the conversations of the flight crew members. The unit is housed in a crash-proof box.

At 2318h, the crew were discussing a "peculiarity" that

had arisen in the readings of two similar pieces of equipment called ADFs (Automatic Direction Finders). These radio receivers were tuned in to a navigation beacon. However, one of the readouts showed that the aircraft was *above* the transmission tower, while the other indicated that they still had some distance to go before arriving overhead. With these identically set instruments giving wildly different computations, the pilot radioed Bali saying that they were *over* the mast and would start their turn toward the airport. At the same time they began descending, and four minutes later reported at a height of 1,675 feet. Nothing else was heard from them. When they had not arrived by the time they had specified, and when they couldn't be raised by the normal communications (Bali Airport was not equipped with location radar), the aircraft was logged as "missing."

As the accident report states: "According to the reconstruction of the flight path based on information obtained from the flight recorders, it is evident that the right-hand turn was made in a position thirty miles north of the beacon . . . " And that was their determination of the "probable causes."

The interesting point, though, is their attack on Pan Americanairways' operations techniques in two "recommendations" attached to the document:

> 1. The Operators should pursue the pilots towards a better knowledge of the Aeronautical Information published in the operations manual for a certain airport, to avoid possibilities of a division of attention during the critical stages of the approach.
>
> 2. A vigil observance by the Operator's Flight Safety Officers [sic] will be appreciable [sic] to well experienced pilots for a possible accident-prone stage within their career [sic], caused by the human factor.

In the case of the crash at Pago Pago, the National Transportation Safety Board (N.T.S.B.) has issued its technical narrative, *NTSB-74-15,* and again, in accordance with the internationally laid-down rules, it contains a paragraph dealing with what they thought happened:

> The National Transportation Safety Board determines that the *probable* [italics added] cause of the accident was the failure of the pilot to correct an excessive rate of descent after the aircraft had passed decision height. The flight crew did not monitor adequately the flight instruments after they had transitioned to the visual portion of an ILS [Instrument Landing System] approach. The flight crew did not detect the increased rate of descent. Lack of crew co-ordination resulted in inadequate altitude call-outs, inadequate instrument cross-checks by the pilot not flying the aircraft, and inadequate procedural monitoring by other flight crew members. Visual illusions produced by the environment may have caused the crew to perceive incorrectly their altitude above the ground and their distance to the airport. The VASI [Visual Approach Slope Indicator system] was available and operating, but apparently was not used by the crew to monitor the approach.

Unfortunately, the National Transportation Safety Board in this case allowed a number of discrepancies to enter their "factual report." The slip-ups became evident after much time was spent studying the background history and documentation from which they had made their findings.

Both at the open inquest, held in Honolulu beginning on March 19, 1974, and in the thousands of pages of information that went into making up their final determinations, there are a large number of references to the weather conditions at the airfield on the night of the accident. No one denies that a raging tropical storm was over the far end of the landing strip at the time the flight was making its approach. No one denies either that the aircraft was *not* in the rain until, at earliest, seconds before the disaster. Yet no one seems to have studied the effects of a heavy shower, its accompaniment with strong, squally winds, nor what would have happened to the plane as it penetrated the edge of the storm.

As Clipper 806 was nearing Pago Pago at 2312h, it will be recalled, the crew asked the approach controller for the latest meteorological details, and the response was "ceiling, estimated at 600 [feet] broken; 4,000 broken; . . . one one thousand, overcast; visibility is one zero [miles]; light rain showers; temperature, seven eight [degrees]; wind, three five zero [degrees] at one seven [miles per hour]; and the altimeter is two nine eight five [inches of mercury]."

Jim Phillips, the co-pilot, then inquired as to the wind's speed and direction. The reply was "variable to the north-northwest here. About three four zero degrees at one seven."

Following that, there was a comment in the cockpit that the air currents were "coming around even more." Whereas just nineteen minutes earlier they had been told that they were from three four zero degrees, they were now being informed that they had changed to zero two zero degrees—a directional shift of forty degrees. And whereas earlier it would have been practically a direct crosswind in the landing area, it was now a left-quartering head wind.

At 2335h, four minutes later, the crew was again advised of the weather conditions at the airport, including "winds, zero one zero [degrees] at one seven, gusting to two three . . . " This indicated to the pilots that the strength had increased a little, but not beyond the forecasted range in the information given them at Auckland.

The next event occurred at 2339h, another four minutes later. The National Weather Service at Pago Pago International Airport made a "special report" on conditions as they were at the time. This was to be radioed to the incoming flight, but it never happened:

> Estimated ceiling, 1,500 feet, broken; 4,000 feet, broken; 11,000 feet, overcast; visibility, 6,000 feet in heavy rain showers; wind, zero four zero degrees at 25 [miles per hour]; altimeter setting, two nine eight five.

While those readings were being prepared, the air traffic controller contacted the flight, then five miles from the landing field, saying he thought the strip's electricity had fused or failed. He could not see the illuminants, but possibly it was because of the heavy rain shower in the vicinity of the building in which he was located. The pilots replied that they had a perfect view of them. They could see the runway lights clearly, straight ahead. This shows that even though the official weather recording gave a maximum visibility range of 6,000 feet *at the airport,* from the plane's position, and between it and the threshold, it was at least 26,500 feet, *if not greater.*

It was then confirmed that there were very heavy showers at the far end of the landing field and the controller passed along the latest information, reading the details off his instruments; "the wind is zero three zero [degrees] at two three [miles per hour] gusting to two nine . . . " This should have indicated to the pilots that the direction had changed slightly once more and now the strength had appreciably increased.

At 2339h56, the windshield wipers started up. These would have been turned on at the very first signs of water on the glass, or possibly even before. (In such a critical phase as the last stages of a landing, the pilots *would not have waited* for a downpour of strength.)

It should be noted that both prior to and after this time, there were numerous confirmations that the runway lights were working and could easily be seen from the Clipper's cockpit. However, there was not one comment about the VASI, which, had it been operating, should have been even brighter than the strip's other luminaries. Nor was there any suggestion made by the survivors that rain was visible on the window panes, though three of them stated that they were looking outside at this stage of the landing. Moreover, before he died, the co-pilot emphasized that up to the time of the impact either none at all or extremely light drizzle only was encountered.

Eighteen seconds after the wipers were turned on, at 2340h14, the flight engineer announced that they were directly above Logotala Hill. And three seconds later, Phillips confirmed that he could see the illuminated runway ahead of them clearly.

At 2340h23, the co-pilot warned that they were "a little high." Immediately afterward the captain presumably increased the rate at which they were descending. The flight recorder at this point, at least, shows a more rapid height loss.

Even at 2340h36, 13 seconds after the 'sink-rate' was deepened, the co-pilot again reported that he could see the airport's lights, further confirmation that the intensity of the rain which could have been falling in the area between them and the landing strip was not great at that time, if, indeed, there was any at all.

Yet five seconds later, at 2340h41, the first scraping noises of the aircraft's underbelly colliding with the top of the Eucalyptus tree could be heard. And forty seconds after that, people who had been on board stated that they were outside the plane in a heavy tropical downpour.

One interesting point, not mentioned in the accident report, is this: If, as the Board says, the pilots were not paying enough attention to their instruments, there is no evidence that they had any difficulty in clearly seeing the airfield and its lights. Yet, three seconds before the scraping noises began, Phillips said "turn to your right."

The Board does not offer an explanation as to *where* Phillips gained the information that they were flying toward the left side of the runway's extended centerline. He could have made the statement only after he had looked at one of two things: Either he was watching out of the windshield at the oncoming strip and noticed that they were heading toward its left—in which case there is proof that the visibility was very clear and discounts the Board's conclusion that there was heavy rain falling at the time, and also that there was "inadequate procedural monitoring by the flight crew members. . . . " Or, he could have been concentrating on his instruments and noticed the incorrect heading on his radio compass or on the flight director—which tends to invalidate their statement that "the crew did not monitor adequately the flight instruments after they had transitioned to the visual portion of . . . [the] . . . approach . . . "

It must be emphasized that the crash-protected recorder does show a deviation in the heading a few seconds before the impact.

The Board also goes on to say that "the VASI was available and operating but apparently was not used by the crew to monitor the approach . . . "

This statement now becomes questionable.

The VASI (the Visual Approach Slope Indicator system) is a series of highly directional lights in two banks on either side near the runway threshold. Each grouping has two sets of beams—one white, the other red. They are very carefully aimed up the incoming flight path at the correct glide angle, giving the pilots a visual reference of the obliquity at which the plane should lose height in order to touch down near the start of the concrete. Naturally, if the pilots misjudged this angle, either they would be too high over the start of the runway, grounding well along it and losing valuable stopping space; or, conversely, they would touch down before the asphalt begins. If they were too high, the VASI would appear to them as white/white. If they were too low, it would be red/red. But if the angle was correct, the lights should be seen as red/white.

One of the major questions is: Why didn't Phillips make some comment about the VASI when he was telling the controller that the runway's illuminations were clearly visible?

The only reasonable explanation seems to be that, because of Logotala Hill, Pan Am had formulated special and specific instructions for Pago Pago. These included the following:

Due to terrain, when landing on runway Zero Five, maintain 1,000 feet [altitude] and disregard VASI until crossing Lima Oscar Golf NDB [that is, the beacon on top of the knoll]. At this point the VASI will indicate "high."

Since the conversations about a possible power failure took place when the Clipper was four and a half miles from the runway (and therefore 2.9 miles before overpassing the navigation transmitter on Logotala) it would appear that the crew was *precisely* following the specified orders.

The Board, in its report, says that the VASI landing-angle aid was "operating and available." Its reason for assuming this is a Federal Aviation Administration document which looks something like a torn-off page from an office "telephone memo" pad. It is headed: "Reminder Memo—avoid errors—put it in writing." And the remainder of the typewritten details are:

> *Date:* February 7, 1974.
> *To:* Whom it may concern.
> *Subject:* Inspection of VASI and MALSR facilities after PAA plane crash, January 31, 1974.
> 02011215Z Arrived site with GFET Takeo Shiraishi and found VASI on Step No. 2 brightness, at 3.4 amps at the substation.
> 02011220Z Arrived MALSR site with (TS) and found MALSR on 10% brightness. Checked 100% brightness with flashers, OPNML. Switched MALSR back to 10% brightness and departed site with (TS). Also noticed VASI lights on while checking MALSR.
> The weather was cloudy but no rain while checking the MALSR.
> GFET's
>
> > [signed] Saboro Fujii
> > Takeo Shiraishi.

The following points should be noted:
1. Even though the heading on the paper used says to

"put it in writing" so as to "avoid errors," this jotting was made *seven days* after the event took place.

2. The N.T.S.B. produced no other proof than this memo that the VASI was operating. (It is quite evident that the notation was made as the result of someone's request.)

3. Neither of the persons who signed the document made further written statements to the investigators. This could indicate that they were not questioned about their claims.

4. If we presume that the information on this "reminder" is correct—that the Main Approach Lighting System Relay was at ten percent brightness, and the VASI was at about two-fifths of its maximum intensity—it could further be assumed that indeed there was no heavy rain (and more than likely no rain at all) in the approaching aircraft's path, again in contrast to the N.T.S.B.'s assumptions.

5. That despite the claims made in the memo, the co-pilot, before he died, emphatically declared to the investigator-in-charge that he did not see the VASI operating while the plane was making its approach. He, in fact, said it THREE TIMES during the interview.

6. In deference to the N.T.S.B.'s findings, the co-pilot also told the investigator-in-charge "there was no heavy rain and no heavy turbulence" during the seconds before the impact. This would mean that he must have had a very clear view of the airport's *operating* lights.

Now, where does all this leave the National Transportation Safety Board and its presumptions that "the VASI was available and operating but apparently was not used by the crew to monitor the approach . . . "?

And if there was a slight "slip-up" in this matter, what about the rest of the Board's suppositions?

Even though it would be impossible to reconstruct exactly the weather conditions that were present at the time the Boeing 707 was making her landing approach, the problems in the vicinity of the flight path can perhaps be understood from a reading of some general studies conducted over the years by the world's weather bureaus.

A few extracts from the British Met. Office's publication, *Handbook of Aviation Meteorology,* a universally recognized authority on the subject, relate to the position Clipper 806 found herself in at Pago Pago that evening and therefore may be appropriate.

First, information on winds:

> There are two distinct types of turbulence or gustiness, distinguished by the terms "frictional" and "thermal." Either type comprises both vertical and horizontal fluctuations of wind, these being practically inseparable. . . . To the occupants of an aircraft, turbulence is recognized as bumpiness and . . . in turbulent conditions the landing and take-off of aircraft may be difficult, for sudden changes of wind speed or direction can cause loss of control when the aircraft is only just airborne.
>
> When the wind is light, [17 miles per hour] or less, and the lapse rate is stable, the airflow over a range of hills is a smooth shallow wave with only feeble vertical currents and no downstream phenomena. . . . With strong winds, vertical currents may be quite extensive and turbulence may be greatly intensified. Frictional eddies form repeatedly in the vicinity of the ridge, break away, and drift downwind before they dissipate. . . . The vertical currents . . . have a noticeable effect on aircraft encountering them. . . . The stationary eddies are to be avoided not only because of the [descending winds] but also because an aircraft encountering the reversal of direction might have its airspeed momentarily reduced below the stalling speed. It is clear that special attention should be given to the wind flow when flights are made near hills and obstructions, and aerodromes should be located well away from hills whenever possible.

In that part of the discussion, only gusty, agitated winds are mentioned. Regarding the Pago Pago crash, according to the position of the Clipper and the crash site, Logotala Hill could more than likely have been producing turbulent eddies around which very strong multi- and contradirectional currents were present.

On turning to the subject of rain, the book emphatically points out that the larger the droplets and the greater their intensity, the more *up currents* there are in the area where the shower is falling. This is because the particles in very light drizzle possess sufficient size and weight to allow them to fall directly to earth. With an updraft present, the droplets are held aloft by the upward-moving currents until a sufficient number of them have collided and combined to create a drop that can penetrate the ascending airstream. Naturally, the greater the speed of the upward moving atmosphere, the necessarily larger size of the raindrop and the more turbulence there is inside the cloud, which forms even greater amounts of rain. Moreover, wherever there are upward streams of air, there also must be, nearby, equal amounts of *down currents.*

> Widespread ascent [upward moving air] occurs mainly within the area covered by a depression (where the cloud takes on a more convective character and precipitation becomes heavy and of a showery nature) and for that reason the connection between low pressure and bad weather is brought about....
>
> As the [descending air] from the rim of the cell nears the ground, it spreads out horizontally and its leading edge has been called the "first gust." The [down current] has been cooled relative to its surroundings so that the outflow above the ground takes on the character of a miniature cold front, often giving a severe squall with a marked increase of windspeed, probably accompanied by a change in direction. The cold air of the down current spreads out all around the cell so that the first gust is usually directed away from the storm; its direction is influenced by the distribution of the several cells of which the storm is composed, but the motion of the storm as a whole tends to make the first gust stronger ahead of the storm than behind it.

When the discussion reverts to both wind and heavy rain combined together, the meteorologists report:

> The cold anafront is the one in which the warm air is ascending and moving forward less rapidly than the frontal surface. Heavy rain usually occurs at the frontal passage, with steady light rain for some time behind the front. It is usually accompanied by a large fall in temperature and a sharp veer and sudden decrease of wind. . . .
> Pressure normally falls somewhat in advance of the front but begins to rise rapidly after the front has passed.

That last point is of extreme interest. Clipper 806's vertical flight path record was found on the crash recorder. The information originally came from the barometric altimeters carried aboard the aircraft itself. It is noted that at 2340h38, the trace shows the height as remaining at approximately 250 feet above the runway for about two seconds. At 2340h40, the readout indicates that the machine suddenly dived or dropped, then collided with the jungle foliage. If the 707, at around 2340h39, passed through the frontal line of the storm, for a few seconds before this the air pressure outside would have fallen and this would have affected the barometric sensors by causing them to show a *false increase* in the aircraft's supposed altitude. Then, as the plane passed through the front, the "rapid rise" in the atmospheric pressure would have registered on the altimeters (and therefore on the recorder) as a very sudden descent. (Barometric pressure increases near the ground, and decreases as height is gained.)

This rise, followed by the fall in recorded altitude and pressure, is clearly indicated on the readout. Under the circumstances, the trace could be explained not as an actual abrupt change in altitude at all, but simply as the airplane passing through the frontal system of the storm. (To back up this point, it will be recalled that no one on board felt such a steep or sudden loss of height as is indicated on the chart—not even the co-pilot. Further, the National Transportation Safety Board makes no attempt at all to explain this extremely peculiar incident.)

Also, when the plane met the turbulent eddies created by Logotala, and the frontal system's "first gust" (both being squally head winds) they would have caused irregular drop-offs in its flight speed. *THIS is visible on the recorder graph.* Countering these reductions in velocity would require constant increases in the engines' thrust output—*which was also done,* as proven by a spectographic analysis of the sounds on the crash-proof dictaphonic device.

Then, at the same time, associated with the frontal system there would have been a strong *down current* of air. This would have caused the Clipper to be gently but steadily pushed into the jungle below, soon after passing through the line-front of the storm. Yet, because of the dropping off of the air pressure, the altimeters could have *incorrectly* shown the airplane to be either losing height at an insufficient rate or, possibly, even to be remaining at the same altitude (as is indicated on the recorder readout)—either of which would have caused the pilot, *who was more than likely working off his instruments because the VASI was not operating,* to think that they were *not* descending fast enough and therefore to slightly increase his angle of decline, unwittingly making the position worse.

And to add to all this, there is also the point that, during the last thirty airborne seconds, the captain a number of times increased the engines' revolutions. These adjustments were evidently made because once the aircraft had penetrated the first gust, the down current of air which would have spread both forwards and backwards on nearing the earth's surface (with most of it feeding the under-storm updraft) would have changed for the pilots from a gusting headwind to a gusting tailwind. This complete reversal of atmospheric direction would have resulted in a rapid and critical erosion of the plane's airspeed.

This too, can be seen, both on the "black box" tape and in the near-to-last conversations on the cockpit voice recorder:

2339h56		[windscreen wipers start]
2340h20	PHILLIPS:	"You're a little high."
2340h25	(UNKNOWN):	"Get your power up."
2340h27		[power throttles advanced]
2340h29	PHILLIPS:	"OK. One-seventy-two [m.p.h.]."
2340h30		[power throttles advanced]
2340h34	PHILLIPS:	"You're at minimums."
2340h35		[speed 169 m.p.h.]
2340h36		[power throttles advanced]
2340h39	PHILLIPS:	"Turn to your right."
2340h41		[impact with trees]

At about 2340h38, Clipper 806 suffered a severe airspeed drop—indicative of a complete change in wind direction. The gusting head winds prior to this would have caused the plane's speed to be quite erratic. But when this altered to a gusting tail wind, a sudden decrement in its flight velocity would have occurred.

The accident report avoids discussing this point as well. Yet all the evidence is there.

The National Transportation Safety Board's findings first became suspect when it accepted the F.A.A. employees' memo that the VASI *was* working, yet evidently *did not accept* the remainder of the same statement which says that the landing lights were only on ten percent brightness. Instead, the Board went into a long dissertation on the visual effects created by heavy rain on a windshield, solely on the basis that the wipers had been turned on and disregarding altogether the co-pilot's insistences that no heavy rain was encountered during the approach. The investigators evidently never asked themselves why the pilots had not requested the lights to be brightened, if indeed they were on only ten percent and the aircraft was immersed in the storm—so that they could be seen more clearly through their supposed water-drenched windshield.

Or, if they were of the opinion that the flyers could see the illuminants clearly, the investigators apparently never questioned the F.A.A. employees' statement that said the switching gear was found set on only ten percent brightness *if,* as they evidently assumed, the aircraft had entered the rain shower at about the time it passed over Logotala Hill. (Not forgetting, of course, that the distance between the air traffic control office and the then invisible runway lights was about 2,000 feet, compared with 9,000 feet between the threshold and Logotala. Yet the controller said he thought the electrics had fused—he could not see them.)

Why would such discrepancies occur in an official report?

It is doubtful whether the National Transportation Safety Board was attempting a cover-up. It would be more reasonable to consider the Federal Aviation Administration's position regarding the crash itself and the circumstances surrounding it.

The F.A.A. was responsible for the equipping, manning, and placement of Pago Pago International Airport and its navigational and ground-to-air aids. It was responsible for "certifying" the landing field, and this meant assessing whether Logotala Hill would, in any way, create a hazard to aircraft movements. It determined it would not. Yet it also agreed with Pan American World Airways when that carrier laid down specific, special, possibly unique operating procedures for its own pilots for landings there—practices which the N.T.S.B. so rightly and severely criticized.

The F.A.A. was aware that there had been numerous power failures at the airport. It did nothing to warn the airlines using the place about these. It even knew that there was a problem in the approach lights' switching system at the very time Clipper 806 crashed: it had a most peculiar electrical fault—one that could not be located. The engineers were fully aware of the fact that pilots had been blinded during the crucial, final moments before touching down, when the lights would suddenly, unexpectedly, jump from low intensity to a full, glaring brightness. (It was because of this that the traffic controller's dimming system had been disconnected.)

It knew that the landing strip was undermanned and underequipped.

The Federal Aviation Administration also knew that the electronic instrument landing system had been misaligned when it was installed; that it perpetually and inexplicably suffered power failures, the causes of which could not be traced; that the angle of the signals was affected not only by heavy rain and Logotala Hill, but also by the many trespassers who frequently roamed around the vicinity of the transmitters. It also knew that, at times, the signals gave completely false readings on cockpit instruments, but did nothing to warn their users.

The F.A.A. was well aware that if anything broke down it took *months* before replacements were obtainable. It knew this, because improper parts were often substituted during the interim period between ordering and receipt. As an example, the voltage regulator for the glide slope landing transmitter was, at the time of the accident, inoperative due to the lack of an integral part. The piece had been ordered on October 14, 1973. It had not been received by January 30, 1974.

The F.A.A. must also have known about unsanctioned and illegal modifications made to certain technical equipment at Pago Pago. During the evening of October 29, 1973, its maintenance technicians lowered the height of the glide slope antenna in defiance of Order 6750.17. Again, on the very day of the accident, an electronics engineer performed an unauthorized adjustment to the same unit. He rewired the transmitter's reset circuit in direct disregard of Order 6032.1. And the same thing was done to the localizer's beacon. (When someone found out about this latter fiddling, instructions were given that the prohibited modification was to be *immediately* removed.)

The Federal Aviation Administration had also had many problems with the electronic gear, especially when the town's voltage supply fluctuated. This happened regularly. The F.A.A. took no steps to warn the airlines using the airport, nor did they instruct the traffic controllers to caution the pilots that the transmitted signals could be unreliable when the power was erratic.

Then, F.A.A. staff members were assigned to the different accident investigation teams, where they worked alongside the N.T.S.B. men. They therefore would have participated in all discussions and findings as the job progressed. Their influence could have been felt.

Pan American, also, was becoming worried. As has been seen, this was the third in a series of four crashes to its Boeing 707 fleet in the relatively brief period of nine months. Of the two that preceded Pago Pago, the Papeete disaster's causes zere completely unknown; but the Boston freight aircraft crash occurred as a direct result of the company's carelessness, as we have already discussed.

Pan Am, likewise, was well represented on all the different investigative groups involved in determining the causes of the Samoan tragedy. Its influence also could have been felt.

However, not all the blame can be leveled against these two implicated concerns.

While the inquiries were still in progress, the National Transportation Safety Board issued one of its well-known safety recommendations. In it, the Board was quick to point out the similarities in three different disasters:

July 10, 1974—Safety Recommendation A-74-55.

On October 28, 1973, Piedmont Air Lines Flight 20, a Boeing 737, was involved in an accident at Greensboro. . . . North Carolina. The flight was attempting a precision approach (ILS) to Runway 14. The accident occurred during darkness, a heavy rain shower and restricted visibility.

Two similar accidents have occurred recently. On November 27, 1973, a Delta Air Lines DC-9-32 was involved in an accident at Chattanooga, Tennessee, and on December 17, 1973, an Iberian Air Lines DC-10-30 was involved in an accident at Logan International Airport, Boston, Massachusetts. Both aircraft were making precision approaches during meteorological conditions that included low ceilings and limited visibility.

Although vertical guidance was provided in each case by an electronic glide-slope, no Visual Approach Slope Indicator (VASI) system was installed for any of the approaches.

If the N.T.S.B. members' findings of the causes of the Pago Pago accident are re-read, it will be quickly noticed that, according to them, Clipper 806 was in *exactly* the same circumstances as the three crashed aircraft mentioned in that recommendation:

1. All indications are that the Samoan VASI was not operating (though if it was, the Board says that the pilots did not use it).
2. They claimed Captain Petersen was landing with the assistance of the electronic glide slope.
3. They asserted that the Boeing 707 was flying through a heavy rain shower when it "bellied" into the jungle.
4. They insisted there was reduced visibility for the pilots at Pago Pago.
5. The Board stated that the accident happened in darkness.

So why was not the Pan American Samoan crash also included in that safety recommendation?

Were they, at the time of its issuance (July 10, 1974), of the opinion that there were no similarities?

Then why did they change their minds between July 1974 and February 24, 1975, when the Pago report was released?

There was an even later safety recommendation dealing with the Iberian Air Lines DC-10 accident at Boston. It was made public on October 3, 1974, as *Number A-74-77,* and it says in part:

> The effect of such a wind shear on the performance of both the aircraft and the flight crew was examined further ... When the situation was reproduced in the simulator, immediate recognition of the wind shear's effects and positive pilot action was required to prevent an impact short of the runway threshold. The pilots who participated in the tests agreed that the restricted visual cues hindered prompt recognition of the developing descent rate and accurate assessment of the pitch attitude change required to arrest descent. ...
>
> A deviation below the glide slope, whether induced by the pilot or by unusual environmental factors, is potentially dangerous during any approach.

From that safety recommendation it was evident that they had made a careful study of the wind's actions on an aircraft that penetrates the periphery of a storm. Yet they did not relate it to the Pan Am 707 at Pago Pago. Why? According to their earlier recommendation, the Iberian DC-10 and the Pan Am 707 accidents were practically identical. Yet in the case of their findings of the causes of the DC-10 affair, it was the wind; and for the 707, it was a whole diatribe explaining how rain creates detrimental visual effects on cockpit windshields and that the pilots were paying insufficient attention both to their instruments and the view outside.

To cap the story, on April 1, 1976, the N.T.S.B. released an unusually thick dossier headed *Safety Recommendations A-76-31 to 44*. The main subject matter concerned the Eastern Air Lines Boeing 727 which crashed at Kennedy Airport on June 24, 1975, killing 113 of those on board.

Only two paragraphs from the document need be quoted:

> The National Transportation Safety Board's investigation of the accident disclosed that the aircraft developed a high descent rate as it passed through or below the base of a mature thunderstorm. The storm was astride the approach course and approximately one mile from the end of the runway. ...
>
> The circumstances of this accident are similar to those of

other accidents which have been investigated by the Safety Board. On May 18, 1972, an Eastern Air Lines Douglas DC-9-31 touched down hard on the runway at Fort Lauderdale, Florida; the airplane was destroyed and three persons were injured. On July 23, 1973, an Ozark Air Lines Fairchild Hiller FH-227B crashed while on a precision approach to the Lambert-St. Louis International Airport, St. Louis, Missouri. Thirty-seven passengers died in that crash. On January 30, 1974, a Pan American World Airways Boeing 707 crashed while on approach to Pago Pago, American Samoa, with 97 deaths. In all these crashes the airplanes were penetrating heavy rain and probably the adverse wind conditions associated with mature thunderstorms.

Why this sudden, belated connection between Clipper 806's casualty and "the adverse wind conditions"?

And why have they *now*—two years after the accident, and one year after releasing their report—decided that the plane was flying in the proximity of a "mature thunderstorm" during its approach to Pago Pago, which knocks out ALL the documentation in their earlier dossier, none of which, even once, mentions the word *thunder,* and adds credence to the suggestions that a large amount of censorship, or cover-up, or just plain fabrication of information seems to have been taking place?

Why are there so many discrepancies in this one inquiry?

Was it because influence was being exerted on the investigators?

Or were the Board's officers really that dumb?

9

The investigation was concerned only with uncovering the reasons why the aircraft ended up where it did, and *that* was as far as it went. What happened to the passengers was only incidental.

So how could such a horrific death toll result from such a relatively gentle accident?

Scientists have known for a long time that if certain everyday substances catch fire, large amounts of combustion products are given off.

Back in 1933, three researchers wrote a paper entitled "Gases from the Thermal Decomposition of Common Combustible Materials." In it, they discussed the emissions of poisonous fumes when wool, silk, and nitrocellulose photographic film were burned. The nitro film, of course, was banned decades ago, but there is now an equally dangerous and universal usage of both manmade filaments and plastics.

As the years passed, because of the dangerous gases created when wool catches fire, the International Wool Secretariat and others tried to evolve a substance that would make it completely unignitable. A number of chemical dips were formulated. None of them was permanent. They could easily be washed out of the fabric in a solution of warm water and soap or detergent. Moreover, the treated woollen materials had to be very carefully dry cleaned, and after ten such operations all the fire retardants had practically disappeared.

Along came the nylons, a boon to war-weary feminine legs in 1945. However their appeal might not have been so universal if more people had realized that this marvelous substitute for silk was, itself, extremely hazardous. Nylon thread in its many grades melts at relatively low temperatures and welds itself to anything nearby—a pair of lovely legs; a hairy male chest; a hand with well-manicured fingers. Melted nylon is nonselective.

After the nylons came the plastics—the polyether foams, the polyvinyl chlorides, the expanded styrenes, the modacrylics—all hailed as the wonders of the modern age. They were, of course, cheaper, easier-to-look-after substitutes for many of the old-fashioned niceties. One of them, polyvinyl chloride, was a replacement for the solid leather on traditional armchairs and lounges. PVC is a cheaper synthetic resin that not only outlasts leather, but is much easier to clean. The horsehair or feather paddings in seats and cushions also gave way to a synthetic: the cheaper, expanded polyether foams. Manmade woolen carpets, too, fell from popularity when the machine-produced, cheaper nylon and modacrylic floor coverings came along.

Aircraft manufacturers quickly took advantage of these products. The original flying machines back in the 1920s and 1930s, had walls that were covered with lightweight plywood. Their seats were unadorned wicker armchairs. The floors, if embellished at all, had small rugs made of wool scattered around the cabin. All this, of course, did little to reduce passenger discomfort caused by the noise and vibrations from the piston engines and the cold from the so-called high altitude.

As the complaints about these problems became louder and more widespread, the airlines and plane makers experimented with methods of soundproofing the cabin. Woolen curtains were added. More and thicker floor coverings were included. And padded and upholstered seats were introduced as a gesture toward providing the passengers with some of the basic comforts.

As civil aviation progressed, fuselage interiors began incorporating a few more features for travelers. There were hat racks, screwed to metal brackets high on the walls. These, like the wall panels, were first made of lightweight wood, covered in an upholstery fabric. Even though they provided an extra convenience and helped reduce noise, they were easily soiled. It was a major task to take them down and clean them.

The invention of the polymetrics then, was a godsend to the industry. Easily formed sheets and foams—the polyethylenes, the rigid and fabric polyvinyl chlorides, acrylonitrile-butadene-styrene, urea-formaldehyde, polyesters, and, of course, the expanded polystyrenes—all these seemed to be exactly what had always been needed for many reasons: their simplicity of manufacture, their low cost, their ease of maintenance.

Soon, aircraft compartments were completely covered with the various manmade materials. Today about ninety percent of the passenger cabin interiors are made of plastic, with approximately five percent wool (representing a percentage mix in the seat covers), and the remaining five percent the metal used in seat and overhead hat-rack construction.

Plastics, plastics everywhere.

The aviation industry was content. Costs had been reduced. Cleaning problems had been greatly diminished.

The enterprises, however, did absolutely nothing to find

out what would happen to the materials in an aircraft fire. It was not as if they were totally unaware of the possible hazards. From the earliest days, long before airline tickets were invented, flying machines had been reduced to molten metal and cinders in accidental infernos. The blazes, it must have been realized, were a direct result, in most cases, of the combustibles being carried. Metal and glass alone do not burn.

Because of this lack of interest in what would happen when modacrylics were exposed to aircraft-type fires, practically no information on the subject exists. Some minor rumblings of concern were heard in America in November, 1965, but that was all. And even though these took the form of pathologists' queries, any recorded papers on this matter have been "lost," evidently because they were not considered to be of much importance. Therefore, in order to learn more about the medical and toxicological effects of fires fed by polymetric acrylics, one has to turn elsewhere for information—to the building industry, for example.

In an unpublished paper, the British Fire Research Station's "Note Number 1025," it is pointed out that the statistics relating to structural infernos from 1955 to 1972 show a difference in the causes of human casualties over the latter part of the survey. Whereas from 1955 to 1959 most house and factory victims were found to have died from either "burns and scalds" or "other and unidentified" reasons; the latter five years, from 1968 to 1972, showed a fourfold escalation in the figures for those dying from the effects of being "overcome by gas and smoke." Moreover, the report also points out that the annual total number of dead in the last five-year period had increased by fifty percent over the previous five years.

Two other changes were also noted:

1. The introduction of plastics into buildings admits the possibility of a wider variety of toxic products in fire gases.

These products may not only be more lethal than carbon monoxide, but might also produce long-term damage in non-fatal casualties.

2. Fire brigade officers report an increase in smoke and noxious fumes associated specifically with the presence of plastic materials in fires in buildings.

As the discussion continues, it will be noticed that a "non-affirmative" language is used in most of the technical reports. Statements such as "could be," "possibly were," "may have," and other "loose" phrases continue to crop up. And the results determined for some tests conducted in different countries on specific plastics arrive at different conclusions.

Originally it was the carbon monoxide gas emitted by the processes of burning that was thought to have been the killer. In the *British Medical Journal* dated November 25, 1972, a brief summary is found of the history of this thinking:

> The toxic effects of carbon monoxide on animal organisms has probably been known by man since the discovery of fire, and we know it was recognized as a dangerous poison in ancient times. It was Claude Bernard who first studied its mode of action, and he showed that blood treated with carbon monoxide was unable to [carry] oxygen. J. S. Haldane is considered as the pioneering investigator of the physiology and toxicology of carbon monoxide. Together with some of his co-workers he performed the now-classic studies of the effects on man of carbon monoxide exposure, and his first paper appeared in 1895. Haldane thought, as a result of his investigations, that the only toxic effect of carbon monoxide was its ability to bind to [blood] at a much higher degree than oxygen, thus displacing oxygen and depriving the blood of its oxygen-transporting abilities. This conclusion was derived from the fact that animals such as mice could live for days with their [blood] completely [saturated] with carbon monoxide if the dissolved oxygen in their [blood] was in-

creased sufficiently by placing the mice in high-pressure chambers. . . .

An important breakthrough in carbon monoxide physiology was made by the Swede, T. Sjøstrand (in 1951), who discovered that carbon monoxide was formed continuously in the human body by the [automatic creation] of hemoglobin. . . . This discovery added to the conception of carbon monoxide as a relatively harmless gas so long as it did not interfere seriously with the oxygen transport of the blood.

Now the question is: Do [carbon monoxide] concentrations of up to twenty percent [in the blood] exert measurable physiological or pathological effects? A few years ago the answer would have been "no," but today it would undoubtedly be "yes." The central nervous system seems to be influenced. This was first shown by McFarland and his associates about forty years ago by demonstrating impaired discrimination of small differences in light intensity at two and four percent [saturation] (McFarland, 1970). Also various performances in tests—for instance, the estimation of time intervals without having a clock, and the duration of auditory signals—are found to be decreased by some investigators at [carbon monoxide] levels of around five percent (Beard and Grandstaff, 1970).

The dates of these latter discoveries should be carefully noted—they were all made in very recent times. Yet, as the article says, fires have been with us for many years.

So when relating the effects of small amounts of carbon monoxide in humans to air accident situations, two of the most important "escape senses," the ability of *seeing* one's getaway route in restricted visibility, and the realization that an evacuation must be made without a *time* delay, are reported as being deranged in even the slightest carbon monoxide-saturated atmosphere.

With universal agreement that smoke is given off as a result of the combustion of products, an article in the *Plastics Institute Journal* of January 1971, has this to say:

> Probably the main hazard of smoke is that it makes escape difficult by reducing visibility. The reduction depends on the

composition and concentration of the smoke, the particle size and its distribution, the nature of the illumination, and the condition of the persons involved. . . .

Assuming that the weight of matter in a smoke is ten percent of the weight of the material that has been consumed, then one pound of material would provide enough smoke to give an optical density of point four [i.e., four-tenths of visibility in clear atmosphere] if mixed in a volume of 14,125 cubic-feet [or approximately half the area inside a DC-10 wide-bodied jet aircraft]. . . . The optical density would result in reduction of visibility to about ten feet, which may be regarded as complete smoke-logging as far as the escape of inhabitants is concerned. . . . Moreover, danger due to smoke has in general been found to precede the danger due to burning or to toxic gases.

Smoke may also incapacitate people physiologically by the irritating effects caused by many of the condensation products. . . .

According to [an industry-wide survey], many of the plastics produce 10 to 250 times as much smoke as wood, and therefore a severe limitation would appear to be called for if these plastics were introduced in buildings.

A closer inspection of the "acceptable methods" of analyzing materials for flammability and combustion products' emissions reveals a major problem. In practically *all* past studies, as irregular as they have been, the so-called scientific standards involve *small pieces* of the fabrics being tested contained in *small experiment chambers* under conditions that would *never* be experienced in real-life fires. Strips no bigger than two square feet are set ablaze with a match, a lighted cigarette, or a Bunsen burner. The progress of the flame up or along the material is then watched and measured, and the results given off by combustion are partly analyzed. ("Partly," because not one of the studies details the full chemical contents; and in only a single test is it noted that the chemists located more than seventy-five different gases in the smoke from burning polyvinyl chloride before giving up. Most other experiments have been conducted to determine the contents of specifically named gases which the analysts *think* may be the more interesting or more harmful to man.) So there is not necessarily any relation between the results of these tests and house and aircraft fires as people have actually experienced them, especially when it is realized that airplane passenger cabins are filled with plastics of all compositions and the conflagrations are usually started by the combustion of large amounts of fuel. (The temperature of flaming kerosene is in the vicinity of 1,790 degrees Fahrenheit.)

Another pronouncement states that the smoke and gases given off by a single modern armchair could, within one or two minutes after ignition, completely fill a three-bedroom house with these noxious substances. A single armchair—yet inside a wide-bodied jet there are up to 500 seats, all containing exactly similar materials.

The synthetic resins, of course, are not as simple as it sounds to test. In America, experiments on polymetrics have proven that the smoke generated differs in its chemical composition when the sample strips are smoldering and when they are on fire. And a further change takes place when the flames causing the ignition are increased in intensity (and therefore temperature). The researchers generally seem to be of the opinion that the higher the source heat, the more toxic the gases. But no one is really sure.

As absolutely no worthwhile work with regard to fires and their effects on humans has been carried out, and as the remains of most victims of building and aircraft infernos are rescued much, much after the fact and usually in a most terrible condition, the pathologists have nearly always tended to quickly determine the causes of death as burning, without caring to try to learn why the person was originally, hypnotically, trapped.

In August, 1974, however, the British Fire Research Station did conduct tests on polyurethane foam. As an aircraft cabin has much more of this material in it than nearly any other plastic, it could be more of a problem in air accidents than in buildings.

The experiments showed that as the temperature of the polyurethane went up with the rapid spread of the flames along the small samples, or were the pieces of foam to be ignited from a hot-fire source, the amount of hydrogen *cyanide* given off proportionately increased, until a maximum of between 508 and 732 parts per million were traceable in the surrounding atmosphere. (One hundred parts per million of hydrogen cyanide is considered fatal for humans.)

> From the records of the concentrations of carbon monoxide and hydrogen cyanide . . . it will be seen that the hydrogen cyanide was evolved rapidly and reached a maximum before the carbon monoxide reached an equivalent level of toxicity. . . . A fire involving such a high load [*sic*] of polyurethane foam clearly presents a considerable danger to persons near the point of origin, both from the rapidity of growth and the high initial toxicity of the gases. . . .
> It is clear already that large loads of [polyvinyl chloride] and polyurethane foam are unacceptable in situations where people may be exposed to fire involving these materials.

So says the building industry. Evidently different rules apply to aircraft.

As far as the more specific problems of airplanes ablaze are concerned, as was said, very little information has been searched out by the civil aviationists and therefore very little is available.

In the VARIG Boeing 707 accident at Paris on July 21, 1973, 123 persons died, more than likely while the machine was still in flight, with a raging fire out of control in the rear two-thirds of the cabin. (By the way, all but one of the fatalities were passengers; all but one of the crew survived.) A stewardess reported that not long before the flames were detected, a woman passenger entered a lavatory at the back of the plane with a lighted cigarette in her hand. She emerged a few minutes later, minus the cigarette. Investigations into the *real* reasons for this disaster showed that the cigarette was placed in the paper-towel waste disposal unit under the sink, igniting the used tissues in it, which in turn caused the plastics of which the container was made (either formed polyvinyl chloride or acrylonitrile-butadene-styrene) to catch fire. The flames shot up the space separating the inner mainly polyvinyl chloride wall from the outer aircraft's metal skin, spreading between them, egged on by the air conditioning, and setting a very large area of polyvinyl chloride ablaze.

There have been other similarly caused fires in toilets. Both Alitalia and Pan Am have lost whole lavatory banks in Boeing 747 Jumbos in this fashion. But all of the remaining cases *to date* have been discovered before they progressed too far, and portable extinguishers have successfully put them out. As a result, the civil air authorities have *asked* (not enforced) the airlines to try to stop passengers from smoking in the latrines. This has been done without regulation and without ordering that the dangerous plastic containers be removed altogether and replaced with nonignitable ones.

But the aviation industry's knowledge of what happens when polymetrics in the cabin ignite date back, at least, to the early Apollo moonship days when the three astronauts were burned to death in a flash-fire during a final rehearsal for their venture into space. These fatalities caused N.A.S.A. to realize the hazards in plastics. They immediately set about financing an enormous program for the development of materials that could be substituted for the synthetic resins—products that would simply not catch fire under any circumstances, and therefore would not give off smoke and highly toxic fumes.

By July, 1970, more than twenty completely nonflammable fabrications had been developed. Possibly the best known of them is *Nomex,* an invention of the huge Du Pont de Nemours company (the creators of nylon) and extensively used for motor racing drivers' suits. Both N.A.S.A. and the racing drivers were quick to see the safety advantages of having noncombustible materials where there was a risk of fire. But the civil aviation industry (even though N.A.S.A., in 1970, described how most aircraft cabin furnishings and fittings could be changed over to the nonflammable inventions) preferred to retain the more dangerous, but slightly cheaper, plastics.

It was not that civil aviation did not realize the hazards. Its awareness of the problems created by the widespread use of synthetic resins in aircraft cabins dates back to November, 1965, and the crash of a United Air Lines Boeing 727 at Salt Lake City. Even though much discussion ensued—mainly by the pathologists who conducted postmortems on the bodies of the forty-three dead passengers—the autopsy notes went "missing," presumably because they had difficulty believing what they had found. In nearly every one of the victims they had discovered small amounts of hydrogen cyanide. The question then was: Why were all these people *eating* cyanide? They did not consider that the substance could have entered the body in any other fashion. And because of the condition of the corpses after being in the fire, they adjudged death to have been caused by burning, without tying in the facts that the interior of the aeroplane was laden with plastics, and most of these give off highly toxic gases when ignited.

A graphic description of the speed at which the flames spread inside the 727's cabin was given to C. Hayden LeRoy, the National Transportation Safety Board's human factors investigator. It is contained in a special report:

> One survivor who related the times of fire progress to his actions, reported that the flames inside the compartment after about ninety seconds extended from the tail forwards to about seat-row seven or eight near the front. . . .
> Five survivors who were seated on the right side of the aft passenger cabin at rows seventeen, eighteen, nineteen, and

twenty, stated that the fire broke out within the first twenty-five to thirty-five seconds after the initial impact. Descriptions ranged from "immediately on first impact" to "one or two seconds after impact." They indicated that "flames erupted between seats 18-CDE and 17-CDE." They "exploded without noise." And "fire came from under the floor and shot backwards and forwards. It was in my seat and all around me. The flames then shot up the right-side wall to the overhead hat-racks."

However, the periphery of the industry was not completely inactive. In July 1966 and again in July 1968, the American Air Line Pilots' Association—the industrial union—in co-operation with a few other semiinterested parties, carried out two special test-burnings of surplus military aircraft dummied up inside to represent civil airliner cabins.

They started the fires by placing containers filled with kerosene right next to the bases of the fuselages. After ignition, they noted that the planes' skin—the normal thirty-mil aluminum—was penetrated in thirty-five seconds. The flames then entered the under-section of the passenger cabin and rapidly spread inside. Using detection probes in the rebuilt compartment, they discovered the first signs of smoke inside just forty seconds after the initial combustion. Ten seconds later, the internal temperature readings taken from stylets placed next to the central cabin ceiling very slowly started to rise. Sixty-five seconds later, the smoke detector located six inches below the ceiling showed visibility saturation. At around the same time, the upper temperature was 194 degrees Fahrenheit, while at other recording stations spread throughout the cabin it was 90 degrees. Two minutes after that, the ceiling temperature had shot up to 570 degrees, yet the probes at seat-top height were recording only 150 degrees.

The tell-tale clues were evident, but they were not grasped then or later. When the upper atmosphere inside the compartment was saturated with toxic gases and smoke, at seat-back height visibility was only six and a half feet, *but* the temperature was well within human tolerance. This classes as nonsense all the pathologists' and coroners' adjudications (and mainly the British) that claim death was due to excessive temperatures and burning. Relating this to the U.K.'s Fire Research Institute's findings, it would be reasonable to suggest that continued life *would have been impossible* within sixty-five seconds of the fires being started because of the concentrations of hydrogen cyanide. This is long before the temperatures themselves would have become critical and would have caused death.

So little is really known about the effects on humans of cyanide gas. The doctors realize that if *any of it* is in the atmosphere, it is *impossible* for a person to hold his breath; moreover, the traces force him to breathe faster and more deeply. This, of course, causes a still more rapid intake of all the toxic products. As most researchers agree that hydrogen cyanide is given off by a large number of the plastics and by wool from the very first moment of ignition, perhaps not much more need be said.

In a lecture given at a specialists' symposium, Bryan Ballantine, a scientist with the British Chemical Defence Establishment, Porton Down, detailed quite fully the pathologists' knowledge to date with regard to hydrogen cyanide:

> Death from cyanides has been a consequence of accidental contamination, suicide or homicidal intent, and judicial execution [in the United States]. With the exception of the execution of criminals, death due to the inhalation of hydrogen cyanide has usually been due to accident, occurring particularly [sic] during fumigation processes. In this respect it is important to remember that, since hydrogen cyanide may be readily absorbed across the skin, respirators may not give complete protection and full protective clothing should be worn.
>
> The inhalation of concentrations of hydrogen cyanide, or the ingestion of massive doses of cyanates, may produce immediate collapse with very few intervening signs. In such [cir-

cumstances which completely stop a person's powers of both sense and motion] the individual may scream, lose consciousness and fall. Sometimes a brief episode of [violent spasms with successive muscular contractions and relaxations] is present. In other cases where there has been exposure to lower concentrations of hydrogen cyanide in the atmosphere, or following the ingestion of cyanide salts, the effects may be more protracted. There may be a latent phase, following which a series of signs and symptoms are forthcoming of which the following are the most frequently recorded . . . headaches, dizziness, constricting sensations in the chest, palpitations, weakness, nausea, vomiting, confusion and coma. . . . Heart rate is often rapid initially, but during subsequent coma a bradycardia may be present which is of sinus origin or due to third-degree heart block. Increased arterial pressure may be found early, but [abnormally high blood pressure] is frequently recorded later. . . .

Convulsions often precede death.

The toxicologist then goes on to point out that once the end has arrived, the cyanide starts to mysteriously disappear from the corpse. Within ten minutes of death, as much as forty percent of the amounts taken or absorbed by the body during the periods of consciousness and coma could have vanished. And after one hour, sixty percent of the traceable concentration has gone. This accounts for the relatively minute figures recorded by the pathologists in air accident victims by the time they actually get around to conducting the autopsies. Yet this point is derided by the airlines and others.

Another aspect, that of the hugely varying percentages of carbon monoxide discovered during post-mortive tests, was explained at the same symposium by D. J. Blackmore of the British Central Research Establishment at Aldermaston: "[Saturation] determinations in accidents associated with fire is of value in assessing the period of time between physical injury and the eventual death of the victim."

Blackmore then showed that in an aircraft crash during

which everyone on board instantly died on impact, the concentrations of carbon monoxide in nonsmoking passengers were between zero and five percent, while that in smokers was from five to ten percent. He then went on to discuss the victims of another disaster in which everyone lived through the landing, only to die minutes later in the post-impact fire. In this case, the average saturation of carbon monoxide was between eighteen and twenty-five percent.

So the rule seems to be: the healthier a person is, the longer respiration will continue after the hydrogen cyanide has taken its rapid comal effects; this is shown by the varying concentrations of carbon monoxide in the body after death. The amounts of hydrogen cyanide discovered during the post-mortem, on the other hand, depend more on the time between extinction and scientific analysis—it decreases with the passing hours.

It is because the aviation pathologists (and mainly the British) have not understood these two points that absolutely nothing has been done about this major problem.

It seems irrefutable, in any event, that Dick Smith and Heather Cann breathed in a mixture that contained hydrogen cyanide and hydrogen chloride after the crash of Clipper 806. Their descriptions of the substance as being "a noxious gas with a slight smell of almonds" adds further, and Smith's likening it to his past experiences with chlorine gas at his Swim Gym more than proves the point.

The woman's scream—a symptom of hydrogen cyanide poisoning—increases the evidence.

At the time of their escape, the fires in the Boeing 707 had been burning for about thirty seconds. (They started at approximately 2340h23; the two escaped at about 2341h11.) This bears out well the theories that the atmosphere would have been uninhabitable *within* forty seconds after ignition, especially since the flames would have had unimpeded entrance to the passenger cabin through the holes torn in the fuselage skin as the machine slithered along the rough jungle floor.

Yet the civil air authorities maintain that there are nearly always ninety clear seconds in which the occupants can get away after a survivable accident, whether a post-crash fire occurs or not. They even legislate for this as being an "acceptable" escape time.

In the Pago Pago holocaust, of the forty-nine bodies subjected to full pathological tests, *the whole lot of them* con tained percentages of these forementioned toxic substances. The amounts varied because of the differing lengths of time they continued to breathe while they lay unconscious in the poisoned atmosphere, waiting helplessly for the end to come. One infant had just four parts per million of cyanide in her bloodstream when the pathologists finally made the tests, with a sixteen percent saturation of carbon monoxide—indicating that her tiny body fell quickly into a coma and moments later died. An athletic adult in his mid-twenties revealed 280 parts per million of cyanide and forty-one percent carbon monoxide, a terribly lethal dose of both—showing that he survived for a longer period, but sense-bereft.

It took only a few lung-fulls of the air inside Clipper 806 for the average person to lose consciousness.

Now surely, one ponders, hadn't the passengers been prepared for an emergency such as the one that occurred that late summer's night in tropical American Samoa?

Weren't they given a briefing before the aircraft departed Auckland on what to do in case of a crash?

Weren't they instructed to take time out to read a pamphlet specially prepared, written, and printed by Pan American World Airways, and inspected and passed as "acceptable" by the Federal Aviation Administration?

The questions become persistent.

Also, didn't all reports of the Clipper's slide through the tropical jungle pointedly remark on the gentleness of that event?

And isn't it right that not one person on board the Boeing 707 other than acting co-pilot James Phillips suffered a

EMERGENCY INSTRUCTIONS
Boeing 707

MESURES DE SECURITE

ANWEISUNGEN FÜR DEN NOTFALL

ISTRUZIONI DI EMERGENZA

INSTRUCCIONES DE EMERGENCIA

INSTRUÇÕES PARA EMERGENCIA

處理緊急事故指導

非常時の場合の心得

PAN AMERICAN
WORLD'S MOST EXPERIENCED AIRLINE

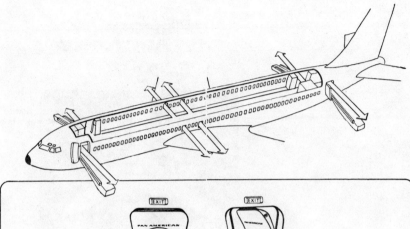

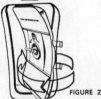

FIGURE 1

FIGURE 2

EMERGENCY DOOR EXITS

The main entrance doors and the galley service doors are emergency exits. There are two exit doors at each end of the cabin. To open — simply lift and rotate the handle fully (FIGURE 1). Push door out (FIGURE 2). Evacuation slides are located at each door for quick egress.

PORTES SERVANT DE SORTIES DE SECOURS

Les portes d'entrée principale et celles situées près de la cuisine sont des sorties de secours. Il y a deux sorties à chaque extrémité de la cabine. Pour les ouvrir, soulever et tourner la poignée complètement. Pousser la porte vers l'extérieur. Des glissières de sortie sont installées à chaque porte pour une évacuation rapide.

TÜRNOTAUSGÄNGE

Die Türen am vorderen und am hinteren Kabinenende und die Bordküchentür sind Notausgänge. Um die Türen zu öffnen, ist der Türhebel anzuheben und dann im Kreis zu drehen. Dann die Tür nach aussen stossen. Notrutschen, die eine schnelle Raumung der Kabine ermöglichen, befinden sich bei jeder Tür.

PUERTAS DE EMERGENCIA

Tanto la puerta principal como la de servicio de proa son salidas de emergencia. Hay dos puertas de salida al final de cada lado de la cabina. Para abrirlas, levantese y girase completamente la manilla. Empujase hacia fuera. La escalera del fondo es también salida de emergencia. Rampas de deslizamiento están situadas en cada puerta para una salida más rápida.

PORTA USCITA DI EMERGENZA

Vi sono altre due uscite in cima e in fondo alla cabina passeggeri. Per aprire, sollevare e far ruotare la maniglia completamente. Spingere il portello in fuori. Per una rapida evacuazione, servirsi dello scivolo di cui e' dotata ogni uscita.

PORTAS PARA SAIDA DE EMERGÊNCIA

Tanto as portas principais como as de serviço para as cozinhas funcionam como saidas de emergência, havendo portanto duas portas de saida em cada extremo da cabine. Para as abrir, basta rodar manípulo toda para cima. Empurrar a porta para fora. Em cada porta existe um deslizador para se poder sair ràpidamente.

緊急出口門

機艙兩端的大門及厨房門都是緊急出口。前後座艙的後部也有緊急出口門。只需將門柄轉一整圈便可打開門，然後將門向外推。每一扇門都附有滑板帮助及加速疏散。

非常口

主要出入口のドアと料理室への通路のドアが非常出口になつています。機内の前後部にそれぞれ二つ非常口があります。ハンドルを唯々上にあげて廻せるだけ廻して（圖１）ドアを押し開けて下さい（圖２）。敏速を要する避難のために、滑り台が、それぞれドアの處に用意されています。

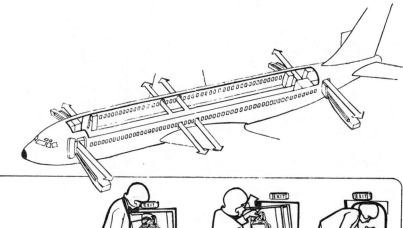

FIGURE 1 FIGURE 2 FIGURE 3

EMERGENCY WINDOW EXITS
Windows marked "Emergency Exit" are located in the middle of the cabin leading onto the wing. To unlock and open, pull handle above the window *inward* (FIGURE 1). Grasp bottom and top of window (FIGURE 2) and discard (FIGURE 3). A life line located in the top of the exit attaches to a ring on the wing for assistance in leaving the aircraft.

HUBLOTS SERVANT DE SORTIES DE SECOURS
Les hublots marqués "Sorties de Secours" sont situés au milieu de la cabine donnant accès aux ailes. Pour déverrouiller et ouvrir le hublot, tirer la poignée placée en haut *vers l'intérieur*. Saisir le bas et le haut du hublot et le retirer. Une corde sera attachée à un anneau au dessus de l'aile pour vous aider à sortir de la cabine.

FENSTERNOTAUSGÄNGE
Die Fenster mit der Bezeichnung "Emergency Exit" befinden sich in der Mitte der Kabine und führen auf die Tragflachen. Zum Öffnen ist der Handgriff über dem Fenster zu ziehen. Oberen und unteren Fensterrand fassen und das Fenster herausnehmen. Ein Notseil befindet sich über dem Ausgang und lässt sich an einem Ring auf der Tragflache befestigen, um so als Halt beim Verlassen des Flugzeuges zu dienen.

VENTANILLAS SALIDA DE EMERGENCIA
Las ventanillas marcadas "Salida de Emergencia" están situadas en el centro de la cabina con salida hacia las alas. Para abrirlas tirase hacia dentro de la manilla situada en la parte superior. Agarrese fuertemente de la parte superior e inferior y echese la puerta hacia un lado. Una cuerda salvavidas situada en la parte superior de la salida de urgencia atada a un anillo en el ala, sirve para facilitar la salida del avión.

FINESTRINO DI EMERGENZA
Ve ne sono due a meta' della cabina in corrispondenza delle ali. Per aprire, tirare la maniglia sopra il finestrino e rimuovere completamente lo sportello con tutte e due le mani. Attaccare la fune di sicurezza situata sopra il finestrino all'anello che si trova sull'ala e servirsene per abbandonare l'aereo.

JANELAS PARA SAIDA DE EMERGÊNCIA
As janelas assinaladas "Emergency Exit" situamse na parte central da cabine, com acesso para cima das asas. Para as destrancar e desalojar, puxar o manipulo localizado sobre a janela. Retirar a janela, pegando-lhe pelas extremidades, e desfazer-se dela. O cabo (corda) de salvação instalado por cima da janela deverá ser ligado a uma argola sobre a asa, a fim de facilitar a saida dos passageiros.

緊急出口窓
標明有『緊急出口』的窗戶是在機艙中側通至機翼處。將窗上的小柄向內拉便可打開窗戶。抓住窗戶的上沿和下沿便可迭離機艙。在出口的頂端有一救生綫繫在機翼上的一鐵環上，可以幫助乘客離開機艙。

非常窓口
『非常出口』は機内の中央にあり主翼の上に向つています。開閉する為めには、窓の上のハンドルを内側に引き（圖1）窓の上下をもつて（圖2）取りはずして下さい（圖3）。救命綱は翼のリングにつながれて出口の上にありますから、それを使つて飛行機から脱出して下さい。

205

EMERGENCY INSTRUCTIONS
PULL OUT BACK COVER FOR ILLUSTRATIONS

1 PUT ON YOUR LIFE JACKET
Your life jacket is located under your seat in a pouch To obtain life jacket pull handle forward When instructed to by Crew—Put arms through loops—then place jacket over head.

THIS JACKET IS DESIGNED TO BE WORN BY EITHER A CHILD OR AN ADULT.

2 SECURING LIFE JACKET
Grasp straps under arms, lean forward in seat and pull back flap down. Pull **YELLOW TAB** on waist straps

3 BRACING POSITION
When crew command says "Brace for impact" lean forward, lower head between knees, clasp arms tightly around legs, and tense muscles for possible landing impact.

4 DEBARK FROM AIRCRAFT
When aircraft stops—not before—unfasten your seat belt. Follow Crew's instructions for debarking from aircraft in a safe orderly manner.

5 LIFE JACKET INFLATION
Only when outside aircraft, pull down sharply on **RED** inflation knobs and the life jacket will inflate automatically. To inflate orally—blow directly into black inflation tubes. **DO NOT INFLATE JACKET INSIDE CABIN.** The jacket light, powered by a water activated battery, can be turned on by pulling the **BLUE TAB** marked, "Pull to Light."

6 EMERGENCY OXYGEN
Oxygen is provided for all passengers. In an emergency the mask for breathing oxygen will automatically become available, at your seat. Pull it towards you, hold it securely over both nose and mouth and **BREATHE NORMALLY.** Oxygen will flow as you breathe. **DO NOT SMOKE.** Extra masks are available for infants. Your crew will advise you when oxygen is no longer required.

7 ESCAPE SLIDE
For rapid escape to the ground, a special slide is installed at each cabin door exit. All crew members know how to use the slides. **OBEY THEIR INSTRUCTIONS.**

8 LIFE RAFTS
Enough rafts are aboard for everyone, each with room for twenty-five people. Every raft contains emergency rations, first aid and water distilling kits and many other useful items. Your highly trained crew will launch the rafts and explain their use.

PUT CHILD'S LEG THROUGH THE STRAP LOOPS SO THAT THE LOOPS ARE BETWEEN THE LEGS.

PULL JACKET ON OVER HEAD. MAKE CERTAIN THE LOOPS ARE STILL BETWEEN THE LEGS.

PULL THE TAB-ENDS AWAY FROM THE CHILD'S BODY TO TIGHTEN THE STRAP LOOPS UP AND BETWEEN THE LEGS.

PLACE TAB STRAP LOOPS AROUND THE WAIST, TIGHTEN AND TIE IN BACK.

INFLATE ONLY ONE TUBE BY PULLING DOWN SHARPLY ON ONE OF THE INFLATION KNOBS.

ANWEISUNGEN FÜR DEN NOTFALL
ILLUSTRATIONEN IM GEFALTETEN RUCKENDECKEL

1 ANGLEGEN DER SCHWIMMWESTE
Ihre Schwimmweste befindet sich unter Ihrem Sitz in einem Beutel. Ziehen Sie die Weste am Handgriff nach vorn heraus. Stecken Sie die Arme durch die Schlaufen, wenn Sie von der Besatzung dazu aufgefordert werden, und streifen Sie die Schwimmweste über den Kopf. DIESE SCHWIMMWESTEN SIND FUR ERWACHSENE UND KINDER.

2 SICHERUNG DER SCHWIMMWESTE
Befreien Sie Ihre Arme von den Schlaufen, beugen Sie sich nach vorn und ziehen Sie die Rucklasche nach unten. Die **GELBEN ENDSTÜCKE** der Hüftbänder ergreifen und nach aussen ziehen bis die Weste fest anliegt.

3 VERHALTEN BEI NOTLANDUNG
Bei der Aufforderung "Brace for impact" (Fertigmachen zum Aufsetzen) beugen Sie sich nach vorn, senken den Kopf zwischen die Knie, klammern die Arme fest um die Beine und spannen die Muskeln für einen möglichen Aufprall bei der Landung

4 VERLASSEN DES FLUGZEUGES
Erst wenn das Flugzeug vollkommen zum Stillstand gekommen ist—und nicht vorher—dürfen Sie die Sicherheitsgurte ablegen. Folgen Sie den Anweisungen der Mannschaft und verlassen Sie die Machine auf geordnete Weise.

5 AUFBLASEN DER SCHWIMMWESTE
Erst nach Verlassen des Flugzeuges—NICHT VORHER—ziehen Sie ruckartig die beiden ROTEN Ventilknöpfe nach unten und die Weste blast sich selbsttatig auf. Die Weste kann auch durch das schwarze Mundstück mit dem Mund aufgeblasen werden. BLASEN SIE DIE WESTE NICHT INNERHALB DER KABINE AUF! Um das Licht an der Weste einzuschalten, das von einer wasseraktivierten Batterie gespeist wird, ziehen Sie an der blauen Lasche mit der Aufschrift "PULL TO LIGHT."

6 SAUERSTOFF-VERSORGUNG
Sauerstoff ist für alle Passagiere vorhanden. In einem Notfall werden Sie an Ihrem Sitz automatisch mit der Sauerstoff-Atemmaske versorgt. Ziehen Sie die Maske zu sich heran, halten Sie sie fest über Nase und Mund und atmen Sie normal. Der Sauerstoff strömt aus während Sie atmen. BITTE NICHT RAUCHEN. Für Kleinkinder, die keinen Sitzplatz innehaben, ist eine extra Maske vorhanden. Die Besatzung wird Sie benachrichtigen, sobald die Sauerstoffmaske nicht mehr brauchen.

7 NOTRUTSCHEN
Zum schnellen Verlassen des Flugzeuges sind an allen Ausgängen besondere Rutschen installiert. Alle Besatzungsmitglieder wissen diese Rutschen zu handhaben. **BEFOLGEN SIE IHRE ANWEISUNGEN.**

8 SCHLAUCHBOOTE
Sind in ausreichender Zahl vorhanden. Jedes Schlauchboot bietet 25 Personen Platz und enthält Notverpflegung, Verbandszeug und Medikamente, eine Trinkwasser-Zubereitungsanlage und weitere nützliche Ausrüstung. Ihre sorgfältigt geschulte Besatzung bringt die Schlauchboote zu Wasser und erklärt Ihnen die Benutzung.

STECKEN SIE DIE BEINE DES KINDES SO DURCH DIE BANDSCHLINGEN, DASS DIESE ZWISCHEN DEN BEINEN LIEGEN.

ZIEHEN SIE DIE WESTE ÜBER DEN KOPF. VERGEWISSERN SIE SICH, DASS DIE BANDSCHLINGEN ZWISCHEN DEN BEINEN LIEGEN.

ZIEHEN SIE DIE BANDENDEN AN UND BEFESTIGEN SIE DIE BANDSCHLINGEN NACH OBEN UND ZWISCHEN DEN BEINEN.

LEGEN SIE DIE ENDEN DER BANDSCHLINGEN UM DIE HÜFTE UND BINDEN SIE DIESELBEN IM RÜCKEN.

BLASEN SIE NUR EINEN DER BEIDEN LUFTSCHLÄUCHE DURCH RUCKARTIGES ZIEHEN AM VENTILKNOPF AUF.

Experienced air travelers normally wear their seat belts fastened while seated even during periods when not required. This is recommended as an added convenience and precaution.

Please do not use any portable radio or TV in flight. These devices interfere with the aircraft electronic system. You may use a hearing aid or portable recorder.

For your safety, please do not use cigarette lighters with clear plastic fuel reservoirs while aboard your aircraft—they may flare up

Normalement, les voyageurs aériens expérimentés maintiennent leur ceinture de sécurité attachée pendant qu'ils sont assis. Même durant le temps de vol lorsque ceci n'est pas requis. Cette pratique est recommandée pour le confort et comme une précaution supplémentaire.

Prière de ne pas utiliser ni récepteur de radio, ni téléviseur portatif pendant le vol. Ces appareils produisent des interférences dans le système electronique de l'avion. Vous pouvez utiliser un appareil auditif ou un magnétophone portatif.

Pour votre sécurité, prière de ne pas utiliser à bord, de briquets munis de réservoir transparent en matière plastique; ils pourraient s'enflammer

Erfahrene Flugreisende sichern im allgemeinen ihren Gurt auch, wenn es nicht vorgeschrieben wird Wir empfehlen dies als zusatzliche Bequemlichkeit und Vorsorge.

Bitte, benutzen Sie an Bord keine Kofferradios oder-TV-Empfänger. Diese Geräte stören das Elektronensystem des Flugzeugs. Hörgeräte oder tragbare Plattenspieler dürfen benutzt werden.

Zu Ihrer eigenen Sicherheit benutzen Sie, bitte, an Bord keine Feuerzeuge mit durchsichtigem Plastikbehälter, da diese aufflammen können.

Esperti viaggiatori tengono di solito la cintura allacciata quando sono seduti anche nei momenti in cui non è richiesto. Questo è raccomandabile come ulteriore precauzione e comodità.

Si prega di non fare uso di apparecchi radio-televisivi. Potrebbero impedire il perfetto funzionamento dell'impianto elettronico di bordo. Sono invece permessi gli apparecchi per migliorare l'udito e i registratori portatili.

Si prega di non fare uso di macchinette accendisigari con serbatoio in plastica. Portrebbero incendiarsi

9243 7477   5 72   LITHO IN U S A

U.S. Federal Aviation regulations require that seat backs be in an upright position on takeoffs and landings.

These regulations also prohibit the carrying on board of any article that cannot be stowed under the passenger's seat. Exempt are such items as a lady's pocketbook or handbag, cameras, binoculars or a reasonable amount of personal clothing.

La réglementation du bureau "U.S. FEDERAL AVIATION" impose que les dossiers des sièges soient en position verticale lors des décollages et atterrissage:

Cette réglementation interdit également le transport en cabine de tout article ne pouvant être entreposé sous le siège du passager. Les seules dérogations concernent les sacs à main, les appareils de photographie, les caméras, les jumelles ou une quantite raisonnable de vêtement.

Auf Anordnung der amerikanischen Luftfahrtbehörde müssen sich die Rückenlehnen der Sitze während des Starts und der Landung in senkrechter Position befinden.

Diese Anordnung untersagt auch, Gepäckstücke mit in die Flugzeugkabine zu nehmen, die nicht unter dem Passagiersitz verstaut werden können. Ausgenommen sind Damenhandtasche, Kamera, Fernglas und einige Kleidungsstücke

I regolamenti dell'Agenzia Federale dell'Aviazione Americana richiedono che lo schienale del sedile sia disposto in posizione verticale durante il decollo e l'atterraggio.

Inoltre si fa divieto di portare a bordo articoli che non possono essere collocati sotto il sedile. Fanno eccezione borsette o borse a mano da signora, macchine fotografiche, binocoli o una quantita' ragionevole di effetti personali.

La reglamentación de la "U.S. Federal Aviation" requiere que los respaldos de las butacas estén en posición vertical para los despegues y los aterrizajes.

Esta reglamentación asímismo prohibe que los señores pasajeros lleven en la cabina, cualquier equipaje de mano que no pusda colocarse debajo de sus butacas. Los siguientes articulos están exentos de esta restricción — bolsos de señora, cámaras, anteojos y abrigos.

Os regulamentos da Aviação Federal dos E.U.A. estabelecem que as costas dos assentos nos aviões devem estar na posição vertical durante as descolagens a na aterragem.

Os mesmos regulamentos proibem que os passageiros levem consigo quaisquer artigos que não possam ser acomodados debaixo da sua cadeira. Não estão incluidos nesta restrição objectos tal como um livro de bolso ou mala de mão de senhora, máquinas fotográficas, binóculos ou roupa de uso pessoal em quantidade razoável.

AWAITING RESCUE
All Pan Am flights are guarded constantly by the Company's Flight Control organization. In addition, the International Civil Aviation Organization provides the means for search and rescue in every corner of the globe. Should an emergency occur these vast facilities are set in motion at once, enlisting all available surface and air transport, both civil and military.

ATTENTE DES SECOURS
Tous les vols de la Pan American sont constamment surveillés par les services de contrôle de la Compagnie. De plus, l'Organisation Internationale de l'Aviation Civile possède des moyens de recherche et de secours dans toutes les parties du globe. En cas d'urgence, tous ces moyens sont immédiatement mis en oeuvre, y compris tour les transports maritimes et aériens disponibles, aussi bien civils que militaires.

BERGUNG
Alle Pan Am Flüge werden ständig durch das Flugkontrollsystem der Gesellschaft überwacht. Darüber hinaus ermöglicht die Internationale Zivile Luftfahrt Organisation (ICAO) Suche und Rettung in allen Teilen der Erde. Sollte eine Notlage entstehen, dann treten diese weltweiten Rettungsmassnahmen sofort in Aktion und ziehen dabei alle verfügbaren Land-, See- und Luftfahrzeuge, gleich ob ziviler oder militärischer Art, zur Hilfeleistung heran.

IN ATTESA DEI SOCCORSI
Tutti i voli della Pan American sono in-interottamente seguiti dall'organizzazione "Controllo Voli" a cura della compagnia stessa. Inoltre l'organizzazione internazionale dell'Aviazione Civile, ICAO, mantiene un servizio costante di vigilanza e di ricerca in ogni angolo del globo. In caso di emergenza, tutti questi dispositivi entrano in azione mentre ad essi si affianca qualsiasi altro mezzo di trasporto aereo o di superficie disponibile, sia civile che militare.

RESCATE
La organización de control de vuelos de la Pan American sigue constantemente el progreso de todos los vuelos de la compañía. Además, la Organización Internacional de Aviación Civil cuenta con medios de busqueda y salvamento en todas partes del mundo. Si llegara a presentarse una emergencia, todos sus servicios se pondrán inmediatamente en acción, utilizando todos los medios disponibles de transporte terrestre, marítimo o aéreo, tanto civil como militar.

AGUARDANDO SOCORROS
Todos os aviões da Pan American estão constantemente acompanhados em vôo pelos serviços de controle aéroe de que esta Companhia dispõe. Além disso a Organização Internacional de Aviação Civil fornece melos para pesquisa e salvamento em qualquer ponto do Globo. Em caso de emergência toda esta vasta organização posta imediatamente em movimento utilizando todos os meios de transporte existentes—de superficie e aereos —tanto civis como militares.

等待援救
所有泛美航機，均由公司方面之飛行管制部門經常嚴密保護。此外國際民航協會有搜尋及救援設備遍達世界任何角落。萬一有緊急事件時，此項龐大救援組織，立即採取行動，民用及軍用之海陸空運輸單位，均儘可能作各種協助救援。

救助機の待機
全パン・アメリカン航空機は、常に社の航空指揮組織で常時監視保護されているのみならず、国際民間航空協会に加入していますから、世界の隅々までに探索・救助ができる体制を整えています。萬一不慮の事故が起きましたときには、この廣大な全機能が直ちに行動を開始し、附近航行中の軍及び民間の船舶或は航空機を即時動員できるよう非常体制を採っておりますから安心して御旅行を楽しんで頂ただけます。

Safety is always our first consideration. These instructions are an added precaution for your personal safety should an emergency ever occur.

REMAIN CALM
Loosen collar and tie, but don't remove clothing. Remove sharp items and high heeled shoes

La sécurité est notre principale préoccupation. Les instructions suivantes ne sont qu'une précaution supplémentaire pour assurer votre sécurité personnelle en cas de nécessité.

RESTER CALME
Dégrafer col et cravate, mais ne pas enlever de vêtement. Oter chaussures à talons hauts et autres objets pointus.

Unsere grösste Aufmerksamkeit gilt der Sicherheit. Diese Anweisungen sind eine zusätzliche Vorsichtsmassnahme für Ihre persönliche Sicherheit, wenn ein nicht vorherzusehender Notfall eintreten sollte.

RUHE BEWAHREN
Lockern Sie Kragen und Krawatte; aber legen Sie keine Kleidung ab. Entledigen Sie sich scharfer Gegenstande und ziehen Sie Schuhe mit hohen Absatzen aus.

La sicurezza è la nostra prima considerazione. Queste istruzioni vogliono quindi essere solo un'ulteriore precauzione per la vostra sicurezza personale in caso di emergenza.

RESTATE CALMI
Sbottonatevi il colletto ed allentate la cravatta senza però togliervi alcun indumento. Liberatevi di oggetti accuminati e di scarpe a tacco alto.

La seguridad es siempre nuestra primera consideración. Estas instrucciones son una precaución mas para su seguridad personal, en caso de que se presentase una emergencia.

CONSERVE LA CALMA
Aflójese el cuello y la corbata pero no se desprenda de la ropa puesta, pero si de todo lo frágil, agudo, punzante y tacones altos.

A segurança é a nossa preocupação principal e estas instruções representam uma precaução que vos garantirá uma maior segurança em caso de emergência.

MANTENHA CALMA
Desaperte o colarinho e a gravata, mas não se dispa. Retire todos os objectos ponteagudos e também os sapatos de salto alto.

安全為本公司服務之第一目標。在有備無患之原則下，如萬一發生緊急事故時，下列指事將是旅客個人安全之額外保障。

齊須鎮靜
解松衣領及領帶但勿脫衣服，其將身上尖硬物品及高跟鞋脫去。

＊安全第一ということは私どもの日切のモットーであります。これらの心得は萬一非常事態が発生したときに備えて慣重を期し、行禄の安全を守るために考えられたものであります。

落着くこと
がもとやネクタイをゆるめて下さい。但し汗衣服は脱がないで、ただナイフのような危険物や御婦人方のハイヒール等は身体からはなしておいて下さい。

209

single broken bone or any other disabling injury in the accident?

Of course it's correct. The answer to every single question is a resounding "YES."

But besides the many already discussed problems with which the passengers on that fateful flight had to contend, there was one other as well. It concerns those Pan Am *Emergency Instructions* mentioned way back at the start of the journey, at Auckland International Airport. And it revolves around the point: *How good are such documents?*

As Pan American World Airways in its Boeing 707 aircraft is still using the same folder today (at the time of writing, well over two years after the catastrophe), it is evident that both the airline and the F.A.A. consider it to be acceptable as far as assisting in the saving of life is concerned.

As Heather and Roger Cann stated when invited to inspect these emergency pamphlets by the crew, and as Dick Smith recalled when he removed the documentation from the seat-back pocket in front of him, Pan American World Airways' escape details are "a complete mess."

They come in eight different languages on the one page. They contain the bare essentials, but they include no statements that would inject into their readers a sense of urgency where speed is needed, or thoughtfulness where a decision of judgment must be made. They omit more than they say.

Since the document is of some public interest—supposedly it was composed and printed and checked solely for the travelers' safety—and, of course, to uphold the minimum requirements of the regulations—parts of it are reproduced in this chapter.

To the untrained eye the instructions, though perhaps somewhat confusing in presentation, seem fair enough. Or do they?

They appear to be about equal to the basic standards that have been accepted for such information all around the world. Or are they?

They say, in rather stilted words, perhaps, roughly the same things that many other airlines try to convey through a series of crude artist's impressions. Don't they?

Then why were NONE of the doors used?

Taking into account the limited details available and speculating on what must have happened inside Clipper 806's cabin, there seems to be only one possible conclusion.

Remembering that it takes a *properly trained* person only a few seconds to move a Boeing 707 door from a "closed and locked" to a "fully opened" position, there was plenty of time for those on board to deploy, at the very least, SOME of the four doors if they had acted quickly enough.

Of the four, the only one remaining after the fire was on the front left-hand side. The investigators found it dangling, partly opened, on its hinges. Firemen and others all confirmed that the door was unlocked, but it was not FULLY opened. The rescuers discovered more than twenty bodies neatly stacked one on top of the other in the immediate vicinity of this egress.

Since NONE of the crew's corpses was found near the doors, what more than likely happened was that as the Clipper slid through the jungle undergrowth, panic-stricken passengers left their seats and headed for the main exits, both at the front and the rear of the plane. This was confirmed by the Canns and Smith. People were out of their places racing for the ways out while the aircraft was still moving. The cabin staff did not take part in this premature escape attempt. Moreover, two stewardesses' bodies were found *well away* from the rear main door. Why they did not lead or at least help with an escape, nobody knows.

As the ground run continued—and quite possibly because it seemed so gentle—others joined those rushing for the exits. This caused a human jam-up around them. Some of the first persons to reach there must have studied the *Emergency Instructions* folders; they managed to get at least one door unlocked.

It seems evident that at the front, passengers were soon pushing and shoving one another in a desperate attempt to get to the exit—thus the reason for someone's calling out: "For God's sake. WILL YOU QUIET DOWN!" It was more than likely the flight engineer, Gerry Green, who shouted this. There were no male stewards on board. The rostered co-pilot, Richard V. Gaines, it will be remembered, had lost his voice. Neither the captain, Leroy Petersen, nor the third officer, James Phillips, went anywhere near the cabin. It *could* have been a passenger who issued the command, but human nature tells us that those at the head of the line, other than a noninvolved person joining them from another direction (such as the cockpit) would, more than likely, have been panicked themselves, and would not have given such an order.

Another point that must be taken into consideration is that in any accident, terrified passengers are *more than likely* to make tracks directly toward either the forward or rear main doors, whichever way they entered the aircraft in the first place, oblivious of the fact that the route takes them directly past a number of other ways of escape.

Angularly opposite the main exits in the Boeing 707 were service doors around which just one or two bodies were found. The concentration was toward the normally used ones. As neither of the service doors was opened, the absolute ineffectiveness of crew seating positions and training standards is evident.

The facts are that the passengers were at the exits before the more than likely completely inactive crew. There was a general panic when difficulty was experienced in getting them opened.

There is a possibility that some doors were jammed, of course. In the case of Clipper 806 this would seem highly unlikely. If they were, questions must certainly be asked about the integrity of the design and its certification by the Federal Aviation Administration, when total jamming could take place under such "gentle" accident circumstances.

But the only one remaining after the fire—the front,

main, left-hand entrance—was *not jammed*. It was found swinging freely on its burned hinges.

No. It is pure fact that the passengers simply did not get the things open in time.

If you know what you are doing, unclosing the exits of a Boeing 707 is really not that problematical. They are "plug-type" in operation. All that has to be done, once the locking handle has been turned, is to PULL THE DOOR INSIDE—*the exact opposite of what the printed emergency instructions say is to be done.*